SYMMETRY AND FUNDAMENTAL PHYSICS: TOM KIBBLE AT 80

SYMMETRY AND FUNDAMENTAL PHYSICS: TOM KIBBLE AT 80

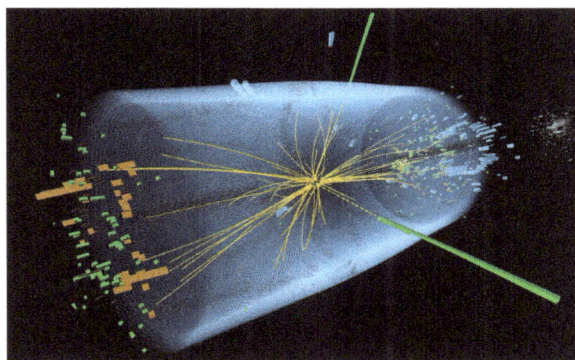

EDITED BY

JEROME GAUNTLETT

Imperial College London, UK

World Scientific

NEW JERSEY · LONDON · SINGAPORE · BEIJING · SHANGHAI · HONG KONG · TAIPEI · CHENNAI

Published by

World Scientific Publishing Co. Pte. Ltd.

5 Toh Tuck Link, Singapore 596224

USA office: 27 Warren Street, Suite 401-402, Hackensack, NJ 07601

UK office: 57 Shelton Street, Covent Garden, London WC2H 9HE

British Library Cataloguing-in-Publication Data
A catalogue record for this book is available from the British Library.

SYMMETRY AND FUNDAMENTAL PHYSICS: TOM KIBBLE AT 80

ISBN 978-981-4583-01-5
ISBN 978-981-4583-85-5 (pbk)

Contents

Preface

Tom Kibble is an inspirational theoretical physicist who has made profound contributions to our understanding of the physical world. To celebrate his 80th birthday, a one-day symposium was held on March 13, 2013 at the Blackett Laboratory, Imperial College, London.

Thomas Walter Bannerman Kibble was born in Madras, India on the 23rd of December 1932. His father, a statistician, was Professor of Mathematics at Madras Christian College. Tom was educated at Doveton-Corrie School, Madras to the age of ten before moving to Edinburgh, where he completed his secondary education at Melville College. His undergraduate and postgraduate education was carried out at the University of Edinburgh from 1951–1958. His PhD thesis on topics in quantum field theory was supervised by John Polkinghorne.

During the final year of his PhD, Tom married Anne Allan and they were happily married until Anne's death in April 2005. They have three children, Helen (born in 1960), Alison (born in 1963) and Robert (born in 1969), and seven grandchildren.

After completing his PhD, Tom spent a year as a Commonwealth Fund Fellow at Caltech. In 1959 Tom returned to the UK taking up a postdoc position with the Theoretical Physics Group at Imperial College, which had recently been founded by Abdus Salam in 1956. In 1961 Tom was appointed to a Lectureship and he then remained at Imperial for the whole of his career. A photograph of the Group in 1964 appears on page **xii**. During his career Tom sustained a major leadership role at Imperial. He was Head of the Theoretical Physics Group for 18 years (from 1971–1983 and 1992–1998) and he was the Head of the Department of Physics from 1983-1991.

Tom's contributions to theoretical physics have received wide recognition. He was made a Fellow of the Royal Society in 1980. He was awarded the Hughes Medal of the Royal Society in 1981 and the Rutherford Medal of the Institute of Physics in 1984 (both shared with Higgs). He was made a Commander of the Order of the British Empire (CBE) in 1998 and was made a Fellow of Imperial College in 2009. In 2012 he won the American Physical Society's Sakurai Prize (shared with Brout, Englert, Guralnik, Hagen and Higgs). In 2012 he was also awarded a Royal Medal of the Royal Society. In 2013 he was made an Honorary Fellow of the Institute of Physics and was also awarded an ICTP Dirac Medal.

Outside of academia Tom has been an active campaigner against nuclear arms. He was Chair of the British Society for Social Responsibility in Science from 1974–1977 and was Chair of Scientists Against Nuclear Arms (SANA) from 1985–1991.

The symposium profiled various aspects of Tom's long scientific career. The tenor of the meeting was set in the first talk given by Neil Turok, director of the Perimeter Institute for Theoretical Physics, who described Tom as "our guru and example". He gave a modern overview of cosmological theories, including a discussion of Tom's pioneering work on how topological defects might have formed in the early universe during symmetry-breaking phase transitions. Wojciech Zurek of Los Alamos National Laboratory continued with this theme, surveying analogous processes within the context of condensed matter systems and explaining the Kibble–Zurek scaling phenomenon. The day's events were concluded by Jim Virdee of Imperial College, who summarized the epic and successful quest of finding the Higgs boson at the Large Hadron Collider at CERN. At the end of the talk, there was a standing ovation for Tom that lasted several minutes.

In the evening, Steven Weinberg gave a keynote presentation to a capacity audience of 700 people. He talked eruditely on symmetry breaking and its role in elementary particle physics. He discussed the role played by the three 1964 papers by François Englert and Robert Brout, by Peter Higgs, and by Gerald Guralnik, Carl Hagen and Tom himself. He also discussed the significant impact of Tom's sole-authored 1967 paper that, among other things, explains the mechanism whereby the W and Z boson get mass while the photon remains massless. Michael Duff of Imperial College gave the vote of thanks.

At the banquet dinner, the UK Minister of Science, David Willetts, stated the importance of fundamental research, praising Tom's contributions. Ed Copeland, of the University of Nottingham and Tom's most prolific collaborator, profiled Tom's scientific leadership, vision and generosity. Tom's son Robert recollected that while his father was doing his amazing work, family life continued as normal — although holiday destinations did strangely seem to coincide with venues for physics conferences. Frank Close of Oxford University concluded the banquet speeches by summarizing the significance of Tom's contributions to the creation of the Standard Model.

Two themes that resonated throughout the day were Tom's extraordinary scientific achievements coupled with his humility.

Jerome Gauntlett,
Head of Theoretical Physics
Imperial College, London
October 2013

Acknowledgments

Many people were involved in helping with the organisation of the symposium. Particular thanks go to several colleagues in the Theory Group at Imperial College: Carlo Contaldi, Michael Duff, Timothy Evans, Jonathan Halliwell, Arttu Rajantie, Kellogg Stelle and Toby Wiseman. Thanks also to Professor Joanna Haigh, the Head of the Department of Physics at Imperial, and Professor Margaret Dallman, the Dean of the Faculty of Natural Sciences at Imperial, for their support. Special thanks to Graziela de Nadai-Sowrey and Katie Weeks who helped in many ways to make the day such a success.

Financial support was provided by the Science and Technology Facilities Council, The Institute of Physics, The London Society of Mathematics, Imperial College and John Hassard, all of which is gratefully acknowledged. Finally, thanks to Phua Kok Khoo and Sun Han of World Scientific Publishing for helping to prepare this volume.

Cover Credit

The cover page (top image) shows an event recorded with the CMS detector at the Large Hadron Collider. It reveals characteristics expected from the decay of a Standard Model Higgs boson into a pair of photons. Image courtesy of McCauley, Thomas; Taylor, Lucas. © 2012 CERN, for the benefit of the CMS Collaboration.

The cover page also includes (bottom image) a simulation of cosmic strings that was obtained using the CSCS supercomputer "Monta Rosa". The image shows cosmic strings in grey and the field density in colour on a plane through the simulation box. Image courtesy of David Daveiro and Jean Favre.

The photo of Professor Tom Kibble is courtesy of Thomas Angus, Imperial College London

Theoretical Physics Group 1964

Front Row: Muneer Rashid, John Charap, Tom Kibble, Abdus Salam, Paul Matthews, Mavis Avis, Ray Streater, Arif-uz-Zaman, Ron King

Back Row: Ansarrudin Sayed, Yahya Khan, Shaun Dunne, Jimmy Boyce, Ghulam Murtaza, Ian Bond, unknown, Ray Rivers, Ian Yamanaka, Ian Poston, Sarwar Razmi, John Strathdee, Ian Ketley, Kamaluddin Ahmad, Dick Roberts

Image courtesy of Blackett Laboratory Photographic Section

Professor Neil Turok; director of the Perimeter Institute of Theoretical Physics, Canada

Professor Wojciech Zurek, Los Alamos National Laboratory, USA

Photos from the symposium courtesy of Meilin Sancho, Imperial College, London

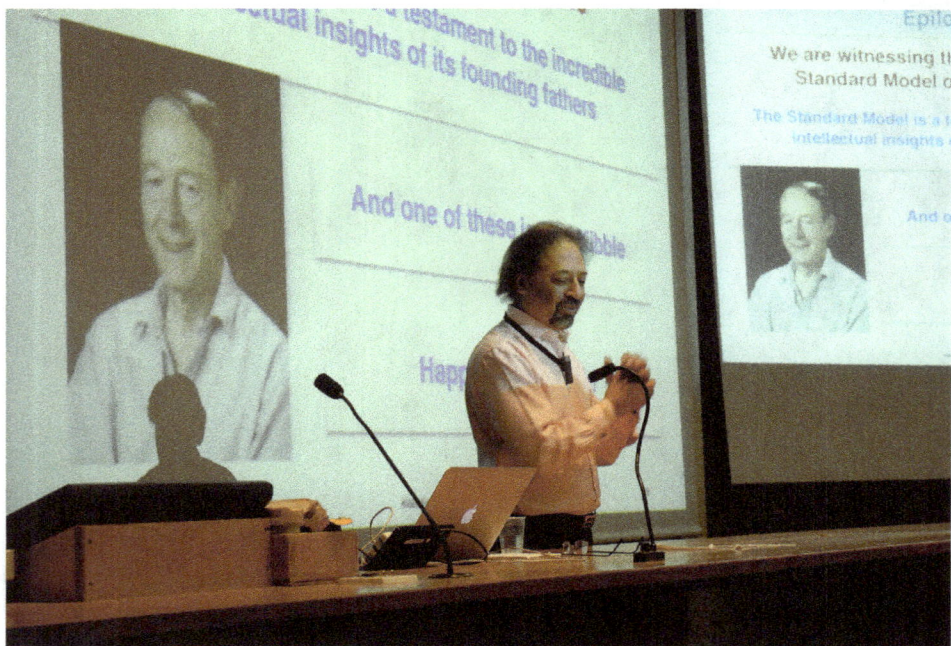

Professor Tejinder S. Virdee, Blackett Laboratory, Imperial College, London

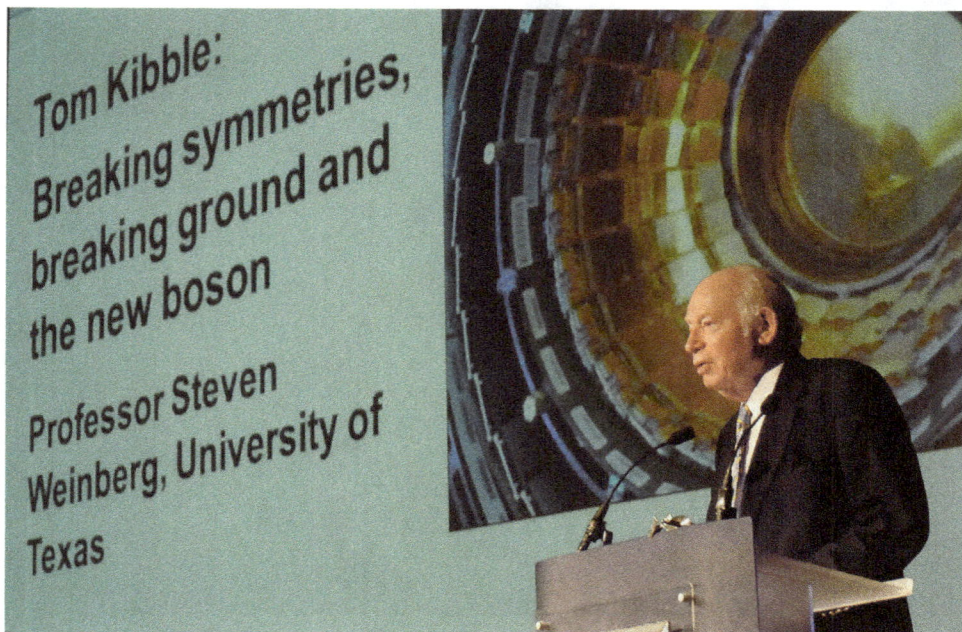

Professor Steven Wienberg, University of Texas at Austin, USA

Audience at the Symposium

(Left to right): T. S. Virdee, J. Hassard, T. Ali, T. Kibble, P. Dornan, J. Pendry, J. Nash

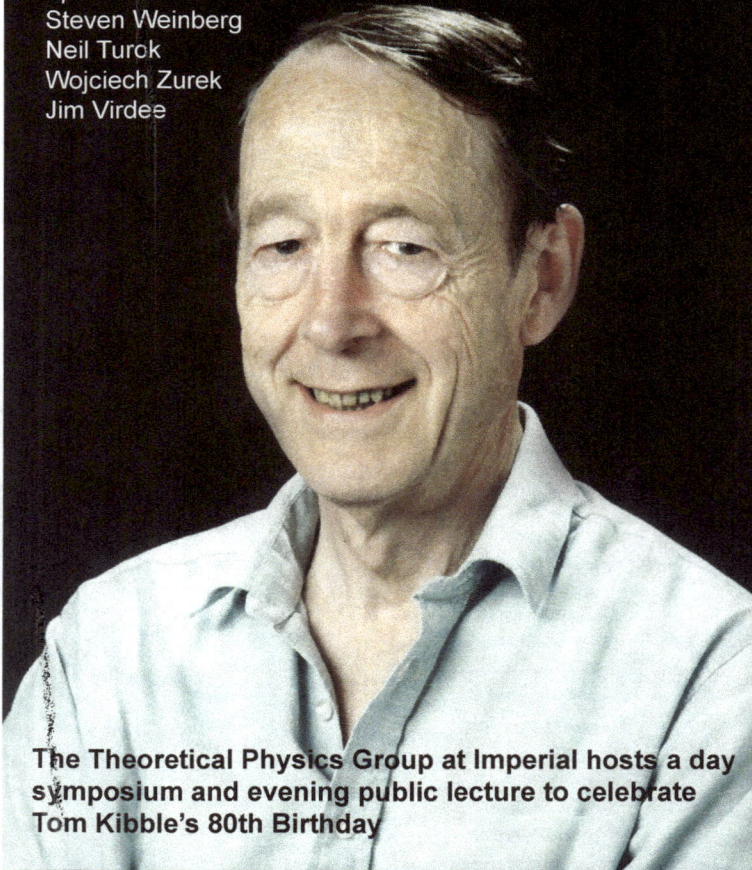

Symmetry and Fundamental Physics
Tom Kibble at 80

Imperial College London, March 13th 2013

http://plato.tp.ph.ic.ac.uk/conferences/Kibble80/

Speakers:
Steven Weinberg
Neil Turok
Wojciech Zurek
Jim Virdee

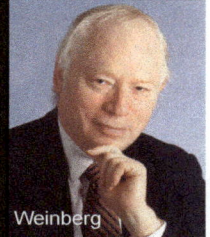
Weinberg

Turok

Zurek

Virdee

The Theoretical Physics Group at Imperial hosts a day symposium and evening public lecture to celebrate Tom Kibble's 80th Birthday

Organisers:
Jerome Gauntlett (Head of Group), Carlo Contaldi, Mike Duff, Tim Evans, Jonathan Halliwell, Arttu Rajantie, Kelly Stelle, Toby Wiseman

Professor Jerome Gauntlett, Imperial College, London
Image courtesy of Imperial College, London

Professor Frank Close, University of Oxford, UK

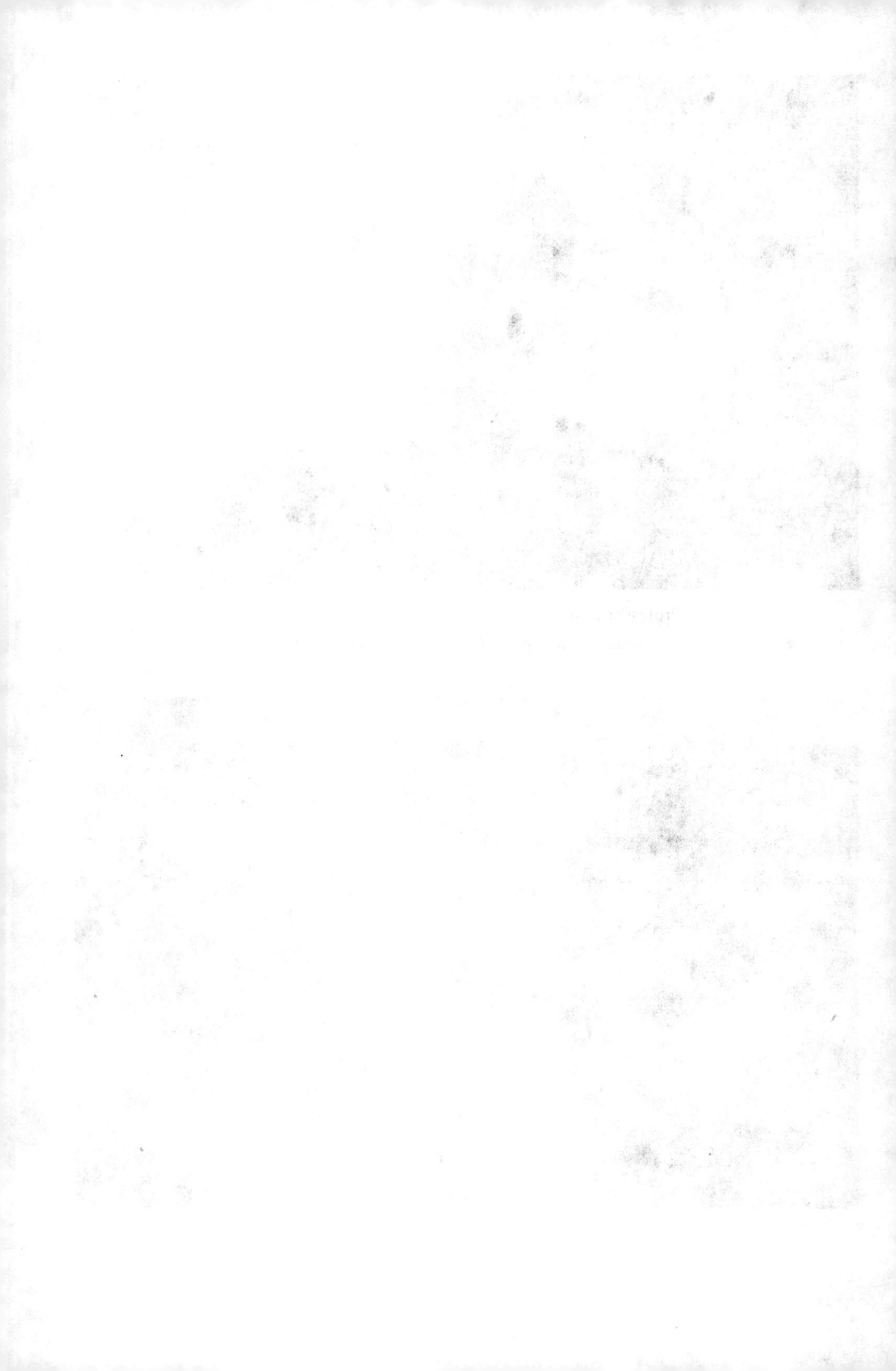

TOM KIBBLE AND THE EARLY UNIVERSE AS THE ULTIMATE HIGH ENERGY EXPERIMENT

NEIL TUROK

Perimeter Institute for Theoretical Physics, Canada

Tom Kibble pioneered the idea that there were one or more symmetry breaking phase transitions in the very early universe, at which defects like monopoles, strings and domain walls would have formed. In the context of grand unified theories, or their extensions, this idea remains compelling: observing these defects would be one of the very few ways of directly confirming the theories. In contrast, inflationary theory invoked a strongly supercooled transition driving a period of exponential expansion which would sweep all such defects away. If inflation terminated slowly, quantum vacuum fluctuations would be amplified and stretched to cosmological scales, forming density variations of just the character required to explain the formation of galaxies. The ensuing paradigm has dominated cosmology for the last three decades. However, basic problems in the scenario remain unresolved. Extreme tuning both of the initial conditions and of the physical laws are required. There are many different versions, each with slightly different predictions. Finally, inflation brought with it the theory of a "multiverse" — a universe containing infinite number of different, infinite, universes — while providing no "measure" or means of calculating the probability of observing any one of them. I will discuss an alternative to inflation, in which the big bang was a bounce from a previous contracting epoch. The discovery of the Higgs boson at the LHC has provided new evidence for such a picture by showing that, within the minimal standard model, our current vacuum is metastable. This opens the door to a cyclic universe scenario in which the electroweak Higgs plays a central role.

1. Introduction

For those of us young cosmologists who were at or around Imperial College (or indeed elsewhere in the UK) in the 1980's and 1990's, Tom Kibble was quite simply our guru. He had realised that grand unified theories were not just appealing theoretical models for particle physics, they would have profound consequences for the cosmos and, most important, they could actually be tested through cosmological observations. This idea of using the big bang as the ultimate experiment, probing the highest possible energies, was and remains extremely attractive.

According to grand unification, the observed pattern of forces and particles becomes much simpler at high energies, becoming unified through a single gauge symmetry principle. At high temperatures, the symmetry would be manifest: there would be only one type of force and only one type of particle. At lower energies, however, the symmetry would be spontaneously broken via the Englert-Brout-Higgs-Guralnik-Hagen-Kibble mechanism,[1] which would be responsible for the variety of forces and particles we observe today. Let me quote from Tom's famous paper of 1976, *The Topology of Cosmic Domains and Strings.*

"In the hot big-bang model, the universe must at one time have exceeded the critical temperature so that initially the symmetry was unbroken. It is natural to enquire whether as it expands and cools it might acquire a domain structure, as in a ferromagnet cooled through its Curie point... The aim of this paper is to discuss the topology and scale of the possible cosmic structures that might arise."[2]

The line of argument Kibble developed led to the prediction of magnetic monopoles, cosmic strings, domain walls and cosmic textures, based on simple and general criteria involving the topology of the degenerate vacuum manifold (Fig. 1). The latter generically depends only on the symmetry group and the sequence of symmetry breaking as the system is cooled. The expansion of the early universe cools the hot plasma within it, and causes the symmetry breaking fields to become progressively correlated on larger and larger scales. The process may be described as a "rapid quench," in which a random pattern of disorder becomes gradually ordered. Kibble pointed out that the causal horizon, set by the speed of light, limits the scale on which any correlations may develop. Such ordering processes are of great interest in condensed matter systems — from liquid crystals to superfluids to superconductors — and topological defects often play a central role. Kibble's picture, developed by Zurek and others, has motivated many experimental and computational studies. More generally, topological defects are of enduring interest in many aspects of quantum condensed matter physics. In fact, one might say defects and ordering processes of the type Kibble considered have been found and studied almost everywhere *except* in the universe! It remains very important to keep looking for cosmic defects, since their observation would provide a unique probe of fundamental physics at extremely high energies.

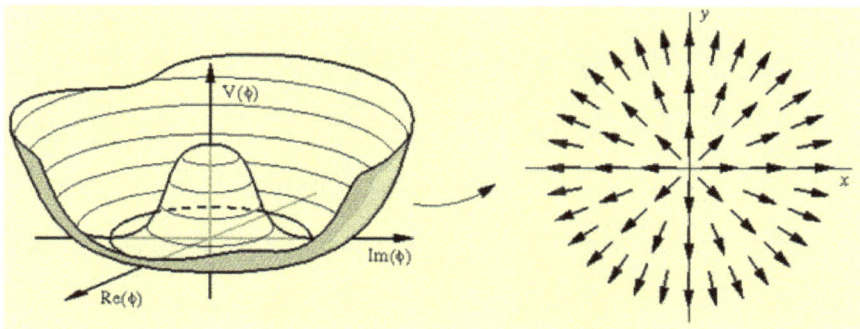

Fig. 1. A topologically stable cosmic string (right), formed as a defect in a Higgs field, whose potential energy (left) has a symmetry-breaking space of degenerate mimima. In the case illustrated, the symmetry is a $U(1)$ phase symmetry and the minimum is a circle.

The general idea is illustrated in Fig 1, in a theory with a single complex Higgs scalar ϕ and a $U(1)$ symmetry under which $\phi \to e^{i\alpha}\phi$. On the left, one has the potential energy per unit volume $V(\phi)$, which depends only on the magnitude of ϕ. Hence, the contours of constant potential energy (and in particular, the space of potential energy minima) are circles in field space. On the right, a configuration of ϕ depicted as an arrow $(Re(\phi), Im(\phi))$ at each point (x, y) of a two-dimensional spatial slice. As one traverses a closed path on this slice, ϕ may wind once around the space of potential energy minima, as shown. If so, then ϕ necessarily vanishes at some point on the slice, and there is necessarily positive potential (and gradient) energy in the vicinity of this point. Such a configuration is topologically stable because the defect in ϕ cannot be removed without changing the topology of the field at large distances. Such a change is impossible without removing ϕ from the space of potential energy minima.

The formation of topological defects depends on the potential energy (more precisely, the free energy) of the Higgs field and how this varies with temperature (Figs. 2 and 3). Figure 2 shows a typical situation for a first order phase transition mediated by a Higgs field Φ, for which a cubic invariant term is allowed. Schematically, the finite temperature effective potential takes the form

$$V(\Phi, T) = \lambda_2\Phi^2 - \lambda_3\Phi^3 + \lambda_4\Phi^4 + Ng^2\Phi^2T^2 \,. \tag{1}$$

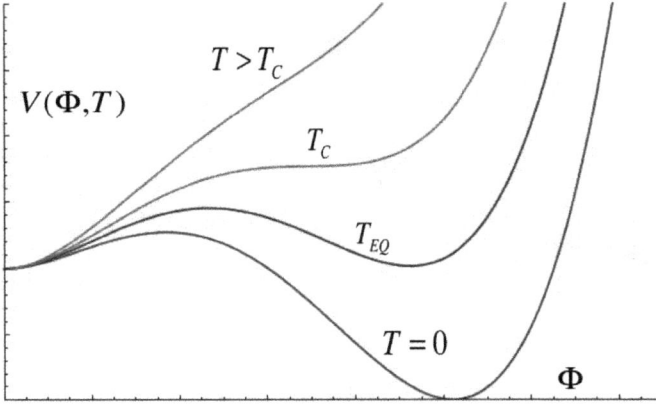

Fig. 2. A first order phase transition, mediated by a Higgs field (or order parameter) Φ. At high temperature, the free energy $V(\Phi)$ is minimised at zero Φ. But as the temperature falls, a new, broken symmetry phase with nonzero Φ appears. Phase coexistence occurs at a temperature T_{EQ}, but as the system is cooled further, the broken symmetry phase dominates. In general, the space of Φ is many-dimensional, and $V(\Phi)$ is constant on orbits of the symmetry group. For simplicity, only one dimension in the space of Φ is shown.

The first three terms are the zero temperature potential, while the last represents the finite temperature correction arising from fields which gain a mass from Φ. It arises because at temperature T, the entropy per unit volume, s, of a massive field degree of freedom with mass m is parametrically lower than that of a massless field by m^2T. The finite temperature effective potential for Φ is determined by the free energy density per unit volume, $F/V = \rho - Ts$, where ρ is the energy density. Equation (1) assumes N degrees of freedom that gain a mass squared of order $g^2\Phi^2$ from Φ. The free energy density may be used to infer the most probable state of the system at temperature T: at high temperature, the symmetric state $\Phi = 0$ is preferred but as the temperature falls below a critical temperature T_C, a new phase with $\Phi \neq 0$ appears. Thermodynamically, F/V is also the negative of the pressure, from which one sees that the two phases can coexist at T_{EQ}, separated by a domain wall in which Φ makes the transition from one phase to the other. As one cools the system, it passes through phase equilibrium before the broken symmetry phase acquires a higher pressure and takes over throughout space.

The typical situation for a second order phase transition is depicted in Fig. 3. When a cubic term is not allowed by the symmetry (for example, when ϕ is a complex scalar field and the symmetry is a phase symmetry), then the finite temperature effective potential takes the form

$$V(\Phi, T) = \lambda(|\phi|^2 - \eta^2)^2 + Ng^2T^2|\phi|^2. \qquad (2)$$

In this case, as the system is cooled below T_C, the unbroken phase with $\phi = 0$ becomes unstable and, in different regions of space, ϕ slides down the potential to lie on the space $|\phi| = \eta$ of broken symmetry minima.

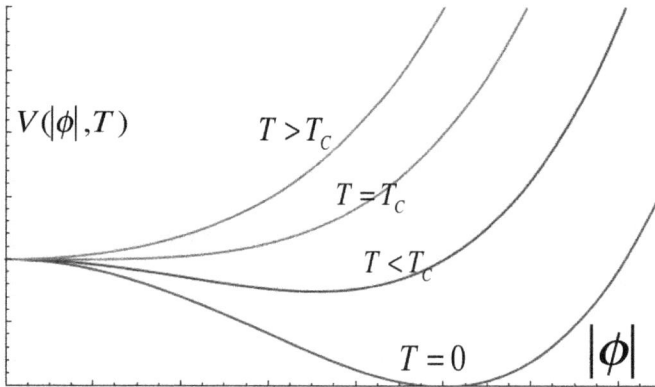

Fig. 3.   A second order phase transition, mediated by a Higgs field ϕ.

The above description, consisting of the tree level scalar potential with the most naïve thermal correction, is oversimplified. When studied more carefully, the breaking of gauge symmetries at finite temperature requires a nonperturbative treatment. Lattice studies of the standard electroweak theory, for example, indicate that for realistic values of the Higgs and top quark mass, the symmetry breaking transition is weaker than second order, and may even be continuous.[4]

In his seminal 1980 paper, *Some Implications of a Cosmological Phase Transition*, Tom began to understand some of the consequences of symmetry breaking in the early universe, especially the formation of defects of various types. He emphasized that bigger questions were at stake, and admitted that his key assumption, that the universe was in a homogeneous state of local thermal equilibrium, at the same temperature T everywhere, could not be justified.

"... the most basic question of all — why the big bang? remains unan-
swered. Moreover, we have no real explanation of the initial state. The
simplest assumption is that soon after the Planck time the universe was in
a state of thermal equilibrium at a temperature not far below the Planck
mass. Why regions that can have had no previous causal contact should
have been in equilibrium is quite unclear, and for the moment must be
taken as an axiom."[3]

2. Inflation

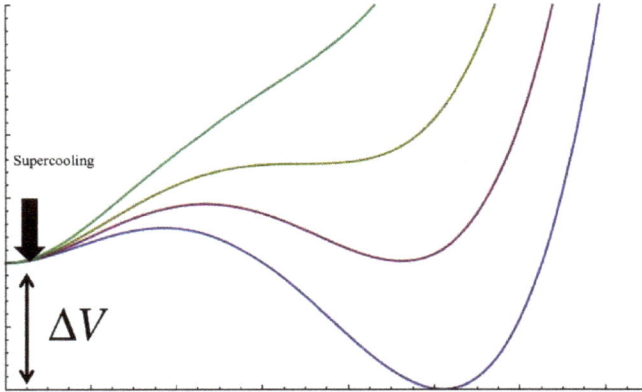

Fig. 4.   A supercooled phase transition, where a scalar field (not necessarily a Higgs
field) becomes stuck in a false vacuum.

Soon after Tom's 1980 paper, Alan Guth wrote his famous paper "Inflationary universe: a possible solution to the horizon and flatness problems."[5]
Guth realised that a strongly supercooled, first order phase transition might
have a dramatic effect on the expansion of the universe. The positive potential energy density associated with the false vacuum corresponds to a
large negative pressure. In Einstein's theory of gravity, negative pressure
is *gravitationally repulsive* and, if sustained, leads to explosive expansion.
The effect may be seen from the Friedmann equation for the scale factor
$a(t)$ (in units where the speed of light is unity):

$$H^2 \equiv \left(\frac{\dot{a}}{a}\right)^2 = \frac{8\pi G}{3}\left(\Delta V + \frac{\rho_m}{a^3} + \frac{\rho_r}{a^4} + \frac{\rho_{aniso}}{a^6}\right) - \frac{k}{a^2}. \qquad (3)$$

The first term on the right-hand side represents the potential energy density ΔV in the false vacuum. It behaves, in effect, like a large, temporary cosmological constant. Subsequent terms represent the spatial curvature, parameterised by k, and the densities of matter, radiation and anisotropy (parameterised respectively by the constants ρ_m, ρ_r and ρ_{aniso}). Assuming that the early universe was expanding, one immediately sees from this equation that ΔV, just because it is a constant, quickly dominates over all the other terms. Once this happens, the solution for $a(t)$ becomes

$$a(t) \sim e^{H_I t}, \qquad H_I = \sqrt{\frac{8\pi G \Delta V}{3}}. \tag{4}$$

It is these two peculiar features: that scalar potential energy (and the associated negative pressure) is gravitationally repulsive, and that it is not diluted away, which are responsible for the exponential expansion of the universe.

Guth argued that an extended period of inflationary expansion, driven by ΔV, would resolve many of the major puzzles about the state of the universe — how it emerged from the big bang so homogeneous, isotropic and flat on large scales, with correlations extending over many horizon volumes. And why the universe possesses such a large entropy when, according to Einstein's equation, a small universe with an entropy of order unity would have also been possible. Finally, Guth noted that a period of exponential expansion would have diluted away any topological defects formed à la Kibble, to negligible densities.

The supercooling scenario is attractive as a robust mechanism for explaining why "pockets" of the universe might get "hung up" and inflated to exponential size. Unfortunately, as Guth himself pointed out, this scenario fails. The problem is that the supercooled phase ends through the nucleation of bubbles, and in general the interior of the bubble takes the form of an infinite negatively curved open universe. Unless the scalar potential is carefully fine tuned, there is generally not enough inflationary expansion within the bubble to flatten the universe out once again. Outside the bubble, the universe continues to inflate exponentially. So one is left with a picture in which vast regions of space are still undergoing super-rapid inflationary expansion, while other parts — the bubble interiors — are nearly empty, with the negative space curvature ($k = -1$) dominating in the Friedmann equation.

The "new" inflationary universe scenario, suggested independently by Linde,[6] and Albrecht and Steinhardt,[7] proposed the following resolution of these problems. Instead of a strongly first order transition, the supercooling

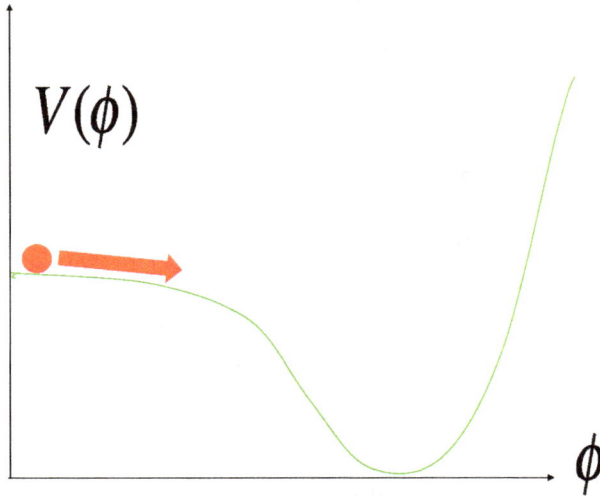

Fig. 5. The "new inflation" scenario, in which a scalar field ϕ slowly rolls down its potential $V(\phi)$ while the universe undergoes quasi-exponential expansion.

would be modest and would end with the inflationary field ϕ being localised near the top of a shallow potential plateau (Fig. 5). From there, it would gently roll downhill, with the universe continuing to expand exponentially until ϕ rolled off the plateau and began to oscillate around the true potential energy minimum. Soon after this improved scenario was proposed, a number of authors pointed out a spectacular consequence. Quantum fluctuations in the inflaton field ϕ — just the usual zero-point fluctuations of every Fourier mode in the vacuum — would be amplified and magnified by the exponential expansion into long wavelength, classical variations in the energy density and spatial curvature of the universe. Thus, new inflation could provide an explanation for the origin of galaxies, based on nothing but quantum mechanics! The picture is enormously appealing in its simplicity and power: a short period of inflation before the hot big bang could explain both to the incredible smoothness and flatness of the universe on very large scales as well as the pattern of small-amplitude density variations needed to seed all the interesting structures on smaller scales.

Following these exciting theoretical developments, and to some extent motivated by them, the ensuing three decades in cosmology were dominated by observational developments. Prime among these was the

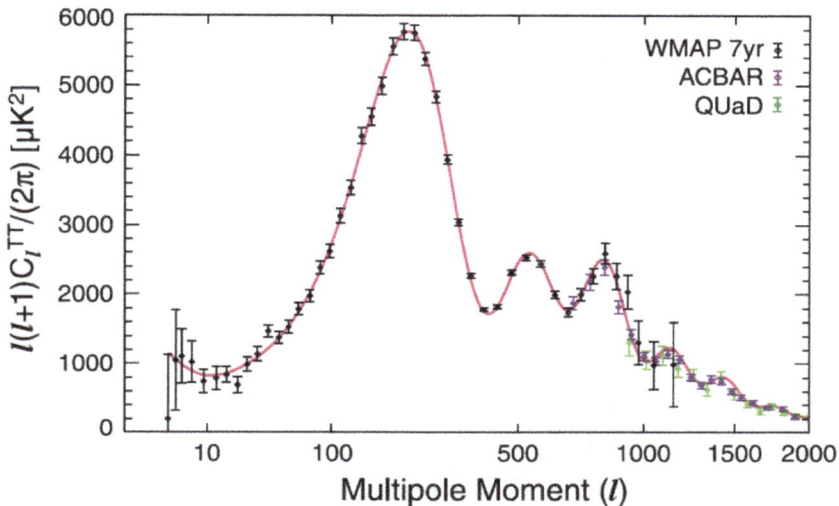

Fig. 6. The angular power spectrum of cosmic microwave anisotropies, as determined from seven years of measurements by the WMAP satellite and other experiments.[8]

progressive mapping of the cosmic microwave sky, which revealed in ever-finer detail the primordial fluctuations as they emerged from the hot big bang. Figure 6 shows the 7 year results of the WMAP experiment,[8] exhibiting the anisotropy power spectrum with exquisite accuracy. The agreement between the theoretical prediction (solid curve) and the observations (error bars) is truly impressive. However, it should be emphasized that the features on the curve are a result of standard physics — specifically, acoustic oscillations (sound waves) in the hot plasma — once very minimal assumptions are made about the parameters of the background universe and the character of the primordial fluctuations. Specifically, it is assumed that the universe is spatially flat, with relative densities of dark energy, cold dark matter, baryons, neutrinos and photons in the proportion of $0.72 : 0.23 : 0.05$, 0.003, 0.0003. Furthermore, the measurements are consistent with the primordial perturbations being: (i) linear (small-amplitude), (ii) growing-mode (scalar, compressional waves), (iii) adiabatic (meaning that the composition of the universe is unperturbed), (iv) very nearly Gaussian and (v) very nearly scale-invariant. As far as fitting the data shown in Fig. 6, only two parameters are required. First, an overall constant representing the dimensionless amplitude of the primordial perturbations and, second, a small tilt representing the fact that the perturbation amplitude appears to increase very slowly with scale. Beyond these two constants, and the requirement of Gaussianity, there is as yet no direct signature of any primordial process.

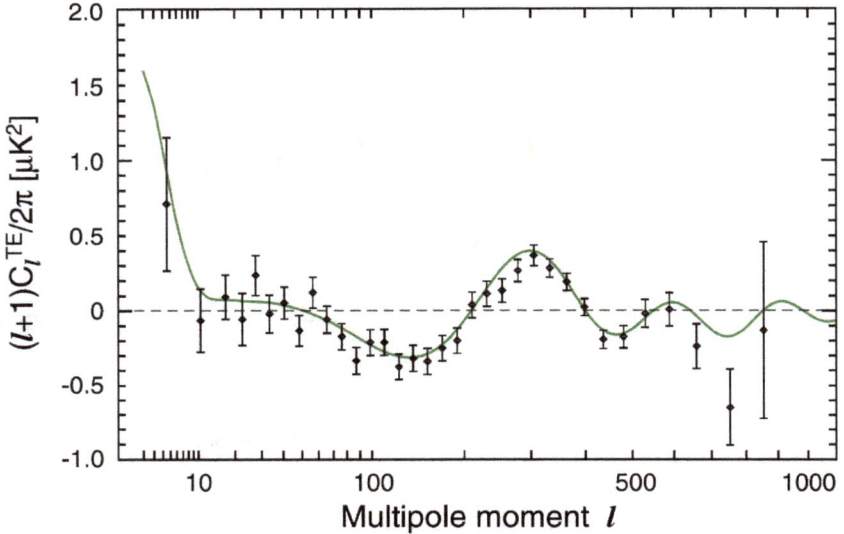

Fig. 7. Power spectrum of temperature-polarization cross correlation of the cosmic microwave sky, as measured from seven years of observations by the WMAP satellite.[8]

Figure 7 provides further confirmation of the incredibly simple nature of the primordial perturbations. It shows the power spectrum of the temperature-polarization cross-correlation, as determined by the WMAP satellite in 2010. This correlation was predicted by Coulson, Crittenden and myself in 1994.[9] There are no additional parameters involved: if the primordial amplitude and the tilt in the power spectrum are fitted to the temperature power spectrum shown in Fig. 6, then the curve in Fig. 7 is an absolute prediction.

Do these observations confirm inflation? They are certainly consistent with inflation but, in my view, there is still room for reasonable doubt. Inflation was invented to resolve the "fine tuning" puzzles of the standard hot big bang — why the universe started out in such a special symmetrical and highly correlated initial state. Yet inflation itself seems to rest on equally serious assumptions about the initial conditions as well as other fine tunings illustrated in Fig. 8.

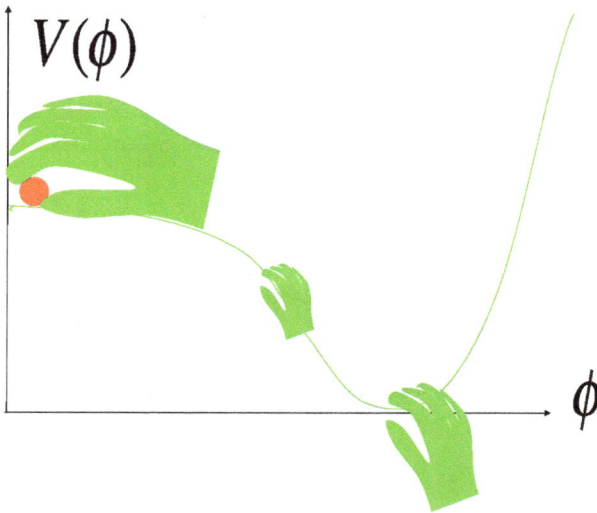

Fig. 8.   Three fine tuning problems of inflation: the unlikely initial conditions, the height
of the potential during slow-roll and the tiny positive value at the minimum.

The first, and most awkward problem for inflation, is the fact that it
requires initial conditions of extremely low entropy. For many physicists,
the beguiling appeal of inflation is that to initiate inflation one only needs a
tiny region with a tiny amount of energy — a few Planck volumes of space
filled with a Planck density of inflationary energy can easily suffice. Once the
exponential expansion takes off, a vast (and potentially unlimited) volume
of space is created. In the process, huge amounts of energy are generated
from a negligible initial amount. As Guth puts it, the universe might be the
"ultimate free lunch." For physicists, used to thinking about problems in
flat spacetime, the energy required is a reasonable measure of how easy or
difficult it is to get something to happen. But in the context of gravity and
cosmology, I believe this intuition is thoroughly misleading. The key point
is that energy is *not* a conserved quantity in an expanding universe. The
relevant conserved quantity is the volume of phase space (or its logarithm,
the entropy), and it is this quantity, *not the energy*, which should be used
to estimate the probability of finding a given initial state.

As an analogue example, imagine walking into a gaming arcade and
seeing a pinball spontaneously fired up into a pinball machine. In principle,
this is perfectly possible. The air molecules surrounding the pinball could,

by chance, happen to collide with the ball and the spring in just such a way as to compress the spring down, before allowing it to drive the pinball up into the machine. But such an event is extremely unlikely. The Boltzmann factor $e^{-E/k_B T}$ for a spring compressed to around one Joule of potential energy at room temperature, is of order $e^{-10^{20}}$. So, in practice, we do not expect to ever see it occur.

Fig. 9. An inflationary initial condition for the universe is like a compressed spring. Just like a spontaneously compressed spring, it is extremely unlikely to find the universe in a high-energy inflationary state if it is selected from a random statistical ensemble.

For a system involving gravity and some form of inflationary energy, there is likewise some probability for finding a small patch of spacetime dominated by inflationary potential energy. Just as with the spring, one would expect the most likely configuration to be a momentarily static patch of space, which had been squashed down to a small size by a previous period of gravitational contraction. The surrounding region of spacetime would resemble the "throat" of de Sitter spacetime, where a universe dominated by inflationary energy bounces from contraction to expansion.

The problem for inflation is that we know how to calculate the probability for such a region, and it is extremely small. The probability is the exponential of the gravitational entropy. The gravitational entropy associated with de Sitter spacetime is just its horizon area in Planck units, M_{Pl}^2/H_{inf}^2 where H_{inf} is the Hubble parameter during inflation and $M_{Pl} = 1/\sqrt{8\pi G}$ is the Planck mass. From the Friedmann equation (3), we find the entropy is $\sim M_{Pl}^4/\Delta V$. For a typical inflationary model, around 100 e-folds or so before the end of inflation, we obtain an entropy of 10^{20} or so. In contrast, the universe around us today has an entropy of around 10^{100} — mostly accounted for by the gravitational entropy of black holes. And the universe we are heading into — an empty de Sitter spacetime with today's dark energy, has a gravitational entropy of $M_{Pl}^2/H_0^2 \sim 10^{120}$, where H_0 is today's Hubble parameter. It follows that the likelihood of inflationary initial conditions, as compared to that for an empty, dark energy-dominated universe, is of order $e^{-10^{120}}$, that is, utterly negligible.

The problem with the unlikelihood of inflationary initial conditions is further illustrated in Fig. 8. The equations of motion for an inflaton field ϕ coupled to gravity form a time-reversible, Hamiltonian system with a conserved (Liouville) phase space density. Calculations of the quantum fluctuations due to inflation assume the slow-roll inflationary trajectory, which is an "attractor" solution going forward in time. However, there are plenty of other solutions, dominated by scalar kinetic energy at early times, which join the attractor going forward. There is no selection principle we know of which rejects these kinetic-dominated solutions over the slow-roll solution. A Liouville-based counting for FRW solutions (which favours inflation, by *assuming* homogeneity and isotropy) indicates that FRW solutions involving N_{inf} or more e-foldings of inflationary expansion comprise an exponentially small proportion, $\sim e^{-3N_{inf}}$, of the available phase space.[10]

A second question mark about inflationary models is that they require a fine-tuned scalar field potential. The height of the potential V during inflation, when our present comoving Hubble volume was swept out of the Hubble radius by the inflationary expansion, must be adjusted to a value smaller than $\sim 10^{-10}$ or so in Planck units, in order to yield the right amplitude for the density perturbations. In fact, more adjustment is now required, because the new Planck data disfavour the simplest one-parameter inflationary models such as monomial potentials (see Fig. 11). All models which remain viable possess at least two adjustable parameters, which are used to fit the two numbers (the amplitude and tilt) required for agreement with the data. This is hardly a compelling confirmation of the theory!

A third challenge for inflation is to explain the vastly greater degree of fine tuning required to fit today's value of the dark energy, of around 10^{-120} in Planck units. Inflation, if it happened, must have been driven by a form of dark energy which dominated at early times and then "turned off" to a precision of a part in a googol. That fine tuning has to be built into any viable inflationary model by hand, an awkward fact which emphasizes our ignorance about the fundamental mechanisms involved.

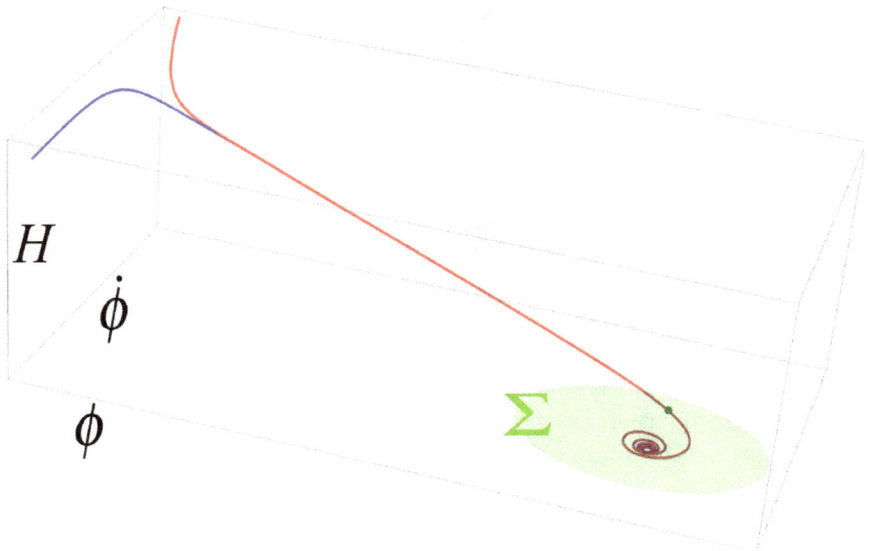

Fig. 10. The slow-roll inflationary solution normally invoked to explain the data such as those shown in Figs. 6 and 7 is only one of a large family of solutions which are generically dominated by scalar field kinetic energy at early times. One can count the number of trajectories giving more than N_{inf} e-folds of inflation by taking a surface Σ in phase space at late times, when inflation ends and the expansion of the universe becomes adiabatic, by using the canonical measure for the scalar field dynamics. The result is that the probability of a solution with more than N_{inf} e-folds of inflation is less than $\sim e^{-3N_{inf}}$, according to the canonical measure. Viable inflationary models require a minimum of $50 - 60$ e-folds, for which the probability is e^{-150} or less.[10]

In summary, slow-roll inflation, with enough e-folds of inflationary expansion to explain the flatness and uniformity of the universe and with the right kind of density fluctuations to explain the formation of galaxies is, as far as we can tell, an exponentially rare and finely tuned phenomenon.

Unfortunately, inflation has even greater problems. The very same quantum fluctuations which are its greatest success, bring with them an ambiguity which, despite decades of effort, has not been resolved. The problem is that even if we just *assume* the inflaton field started out in the right, inflating state, quantum fluctuations in the field as it rolled downhill towards the true vacuum would kick the field back upwards occasionally, causing some volumes of the universe to undergo even more inflation. According to stochastic models of this process, a state of eternal inflation would result, in which the vast majority of spacetime would continue to inflate whilst

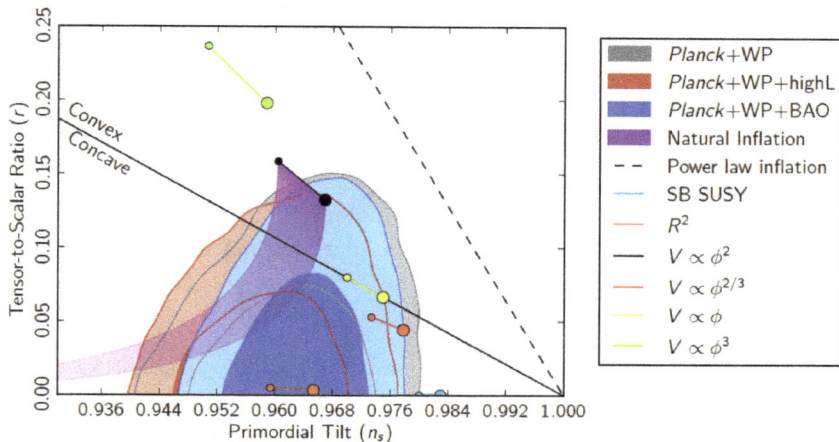

Fig. 11. Inflationary models compared to the measurements of the tilt of the scalar perturbation spectrum (horizontal axis, with exact scale invariance corresponding to $n_s = 1$) and the relative contribution (r) of the gravitational wave (tensor) perturbations. Note that the original ϕ^4 "chaotic inflation" model is completely ruled out due to its overproduction of gravitational waves, and the simple ϕ^2 model is disfavoured by around two sigma. The other plotted models require further specification of the potential (and parameters) to create a minimum. The one model shown which fits the data well, the R^2 model originally proposed by Starobinsky,[11] is questionable theoretically since it involves higher derivative corrections driving the inflationary instability and hence may be compared to the run-away electron problem in electrodynamics.

spawning "pocket universes" in which inflation would terminate and standard hot big bang evolution would take place. The great problem with this picture is that there is no known measure from which one could calculate the likely state of a pocket universe — how curved it is, how many e-folds of inflation it underwent, and so on. It is ironical that inflation, whose main motivation was to render moot the measure on initial conditions for the universe, now suffers from a debilitating measure problem.

These huge conceptual problems of inflation — in particular the fine tuning puzzle and the measure problem — have driven many to resort to the anthropic principle: the idea that the universe is the way it is because we are here. If, in the region of the universe which surrounds us, the dark energy density was any greater, or the inflationary fluctuations any larger, or if inflation had run for too few e-foldings, the argument runs, then the region around us would have been inhospitable to life and we would not be here to ask the question. Whilst it seems perfectly reasonable to correlate our existence with the properties of the universe around us, the absence of a measure makes this difficult to do in any reliable way. At a gut level, I cannot believe that the astonishing *simplicity* of the universe on large scales — the extraordinary uniformity, the near-scale invariance and near-Gaussianity of the fluctuations — are consequences of selection for our own evolution. It would be a case of the tail wagging the dog. The evidence points to a universe of extreme simplicity, far greater than needed for our existence. It is our task as physicists to understand how such simplicity and regularity emerged as a consequence of underlying mathematical laws.

3. Can We Do Better?

Inflation isn't so much a theory as a model-building paradigm. As such, it has been extremely successful — a vast number of models have been proposed, with many variations and adjustable parameters. A vast, self-sustaining literature has been created. However, the basic idea, that there exists a dynamical attractor capable of washing away all of the details of the prior universe and inevitably creating a universe like ours, rests on a naive, intuitive "prior measure" which, as I have discussed above, disagrees with calculations of gravitational entropy and phase space arguments, and has never been otherwise substantiated. Ignoring these problems, model-builders just *assume* the universe starts out inflating, with initial conditions chosen to give a long epoch of slow roll inflation. And all attempts to make sense of "eternal'" inflation have so far foundered. As Paul Steinhardt has recently argued, the framework for inflation is too generous, the models are too numerous, the remaining pitfalls too big and the predictions too broad for us to consider inflation as a testable scientific hypothesis.[12]

Having said this, it is remarkably difficult to do better. Many competing explanations of the formation of cosmic structure — such as cosmic strings and textures — have been decisively ruled out.[13] But the idea that the big bang was a bounce, which underlies the ekpyrotic/cyclic theory and related scenarios, remains a viable alternative.[14]

If the big bang was a bounce, preceded by another large, smooth universe like ours, then many of the problems which inflation was invented to solve are naturally resolved, *without* inflation. For example, there is no horizon problem: regions on opposite sides of today's visible universe were causally connected within a tiny region in the pre-big bang phase. And it turns out to be just as easy to establish flatness, homogeneity and isotropy in a collapsing universe as it is in an expanding inflationary one. Likewise, there are mechanisms available for growing quantum fluctuations into nearly scale-free, nearly Gaussian, density fluctuations to seed cosmic structure formation. Finally, a cyclic universe becomes possible, in which there is no need to postulate a 'beginning,' because time can run back arbitrarily far into the past.[15]

For example, consider the ekpyrotic/cyclic scenario, in which it is assumed that the big bang was preceded by a phase of very slow contraction. Equation (3) is replaced by

$$
\left(\frac{\dot{a}}{a}\right)^2 = \frac{8\pi G}{3}\left(\Delta V + \frac{\rho_m}{a^3} + \frac{\rho_r}{a^4} + \frac{\rho_{aniso}}{a^6}\right) - \frac{k}{a^2} + \frac{8\pi G}{3}\left(\frac{\rho_{ek}}{a^{3(1+w)}}\right), \quad (5)
$$

where in the last term $w \gg 1$. Such an equation of state describes a slowly rolling scalar field with a steep negative potential (for example, $V(\phi) = -V_0 e^{-c\phi/M_{Pl}}$ with $c \gg 1$). In a slowly contracting universe, the ekpyrotic energy density ρ_{ek} would rapidly dominate over all other forms of energy, and spatial curvature. Just as in inflation, since the scalar field energy density is homogeneous and isotropic, the universe would also become homogeneous and isotropic, but this time as a result of slow contraction rather than exponential expansion. During this epoch of slow contraction, quantum fluctuations in the ekpyrotic scalar field are amplified and acquire a scale-invariant spectrum. Again, this is closely analogous to the production of scale-invariant perturbations during inflation. The simplest ekpyrotic models generate a slightly red scalar spectrum, just as the simplest, monomial inflationary models like ϕ^2 inflation.[16]

But there is one very important difference: the ekpyrotic models, operating through slow contraction rather than rapid expansion, do not generate a significant level of gravitational waves. Comparing with the recent Planck observations, shown in Fig. 11, the simplest ekpyrotic models are close to the centre of the favoured region (on the $r = 0$ line) whereas the simplest (monomial) inflationary models like ϕ^2 or ϕ^4 are now disfavoured. It will be fascinating to see the constraints become more precise within the next couple of years.

Generating the scale-invariant perturbations during a period of contraction, rather than expansion, has another important consequence. Regions in which the scalar field fluctuates upwards on its downhill career, delaying the hot big bang, just continue shrinking for longer. This is the opposite of what is found for inflation, in which regions where reheating is delayed undergo exponentially more expansion and therefore dominate the spacetime volume. In an ekpyrotic scenario, most of the spacetime volume of the universe would look the same: there is no multiverse problem.

4. The Electroweak Higgs: A New Clue

Within the past year, an important new clue has emerged. The discovery of the electroweak Higgs particle at the Large Hadron Collider, described by Virdee in this volume, was a *spectacular* confirmation of the mechanism of spontaneous symmetry breaking developed by Tom Kibble and others, a mechanism central to the standard model. The fact that no superpartners or other weak scale particles have yet been found suggests that the minimal standard model may actually be a complete description of nature up to very high energy scales. The main evidence against such a view is *naturalness*: that it is hard to explain why the weak scale is so much smaller than the Planck scale without invoking some symmetry, like low-energy supersymmetry, to protect it. But supersymmetry, while mathematically appealing, brings with it a plethora of unobserved particles and fields, and yet more free parameters. It is worth looking for other, more economical, deep symmetry principles which might explain not only the weak naturalness problem but also shed light on the (far more severe) naturalness puzzle of the crazily small value of the dark energy density in Planck units.

In any case, until there is evidence to the contrary, the minimal hypothesis deserves to be taken most seriously. The standard model is consistent and complete as a quantum field theory, and can be reliably extrapolated to very high energy scales using standard perturbation theory and the renormalization group. Based on the experimental measurement of the Higgs mass, the effective potential for the Higgs field can be computed by running the standard model couplings, with the role of the energy scale played by the Higgs vacuum expectation value (or "vev"). The result[17] is shown in Fig. 12.

The surprising result is that the minimum we live in, in which the electroweak Higgs vev is ~ 250 GeV, is *not* the lowest potential energy state. At nearly 3-σ confidence (see Fig. 13), our vacuum is metastable. Its lifetime is long — of order 10^{800} years — but finite! This means that the dark energy phase we are now entering, while extremely long-lived, will eventually terminate with the nucleation of a bubble of the lower energy phase. This nucleation would spark the creation of a phase similar to that postulated in the ekpyrotic scenario, ending in a 'big crunch' at which the Higgs field would reach Planckian values.

If we take the Higgs potential shown in Fig. 12 seriously, as I have argued we should, it poses yet another problem for inflationary cosmology. Not only do we have to explain why the inflationary field contributes a huge positive vacuum energy during inflation, we have to explain why the

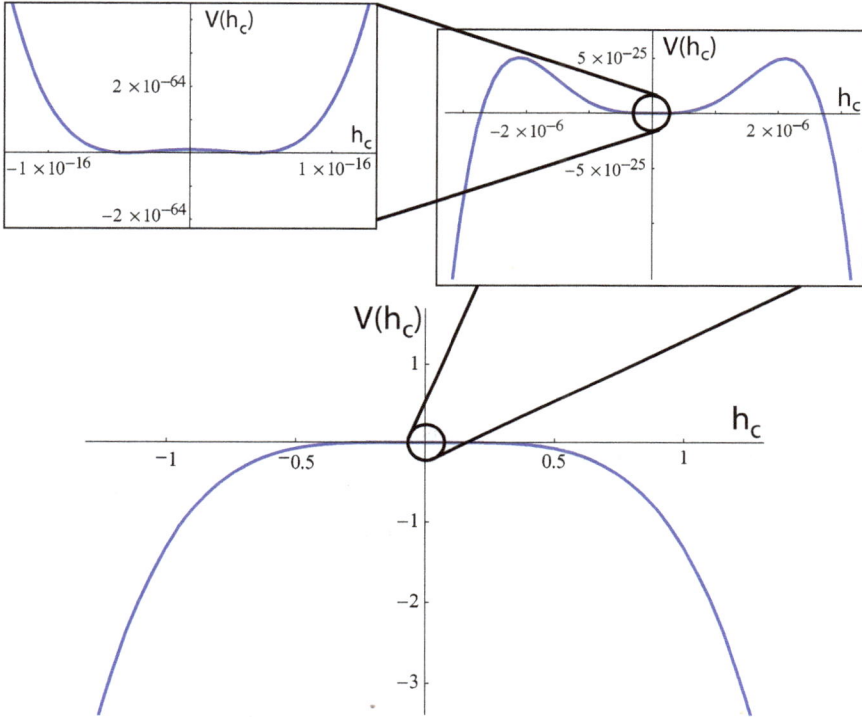

Fig. 12. The peculiar shape of the effective potential V in the standard model, shown as a function of the Higgs vev h_c, in Planck units.[17] Today, h_c lies at the bottom of a tiny well, $h_c < 10^{-16}$, with $V \sim 10^{-120}$ accounting for today's dark energy density. As h_c increases, the potential V rises to a maximum at $h_c \sim 2 \times 10^{-6}$ beyond which it plummets to large negative values. As h_c rises to unity, the potential becomes well approximated by λh_c^4 with λ a small negative constant.

electroweak Higgs field remained localised inside the 'well' at very small values of h_c. If the universe started out at the Planck temperature, one would expect typical fluctuations in h_c of that order, and these would have easily kicked it out of the well. The cooling rate H divided by the interaction rate Γ may be estimated at $\sim T/(M_{Pl}\alpha^2\sqrt{N})$, where T is the temperature, α a typical gauge coupling, and N the number of relativistic degrees of freedom. For temperatures close to M_{Pl} the cooling rate is much larger than the thermalisation rate. Hence there would be no time for thermal effects to localise the Higgs field. So, we conclude the initial value for the electroweak Higgs has somehow to be tuned to $< 10^{-6}$ in Planck units

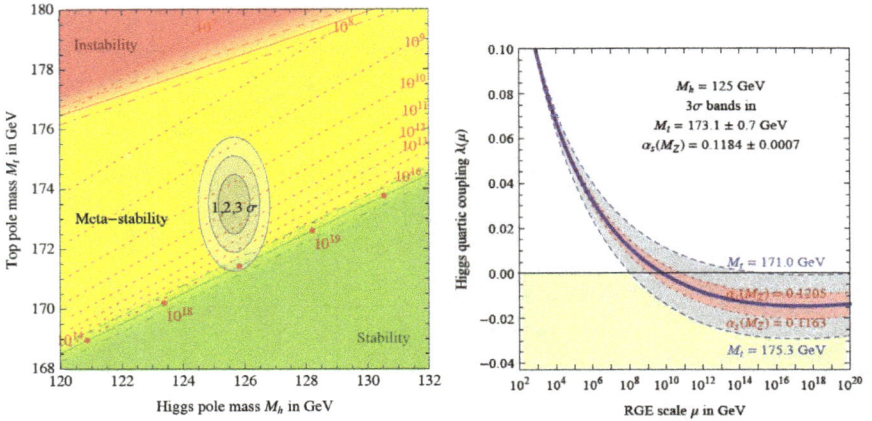

Fig. 13.   Implications of the LHC measurements of the electroweak Higgs mass. The left plot shows the nature of the current electroweak vacuum, as compared to the experimental bounds on the Higgs and top mass (the two most relevant parameters). With high confidence, our vacuum is metastable. The dotted red curves show the energy scale at which the Higgs quartic coupling runs negative. The right hand plot shows the running of the Higgs quartic coupling with energy, showing the intriguing feature that the coupling has a minumum near the Planck scale.[17]

over the region which started out inflating, and thermal effects have to be somehow ignored. Similarly, one has also to explain why the inflation did not quickly terminate by the nucleation of a bubble of negative Higgs potential energy inside the inflating region.

In contrast, the implied metastability of the electroweak Higgs fits the cyclic universe picture to a tee.[15] A cyclic universe requires the future dark energy phase to terminate, and this naturally occurs through the nucleation of a bubble. Inside the bubble will be an ekpyrotic phase, leading to big crunch and, presumably, the next big bang. The key challenge for such a picture is describing the bounce from crunch to bang, a topic I will turn to now.

5. Weyl Invariance and the Big Crunch/Big Bang Transition

In recent work, Itzhak Bars, Paul Steinhardt and I have explored the description of a crunch-bang transition within the context of low-energy effective theory. At first sight, such a description might appear problematic since we necessarily have to pass through a region of Planckian curvatures (in Einstein frame), where at the very least one would expect higher derivative corrections to be important. However, such a view may be too pessimistic, specially if one takes seriously the idea that the laws of physics have an exact underlying conformal symmetry. If so, then one is free to change to a conformal frame in which the metric is not singular at all. In fact, it is not hard to show that the entire standard model can be 'lifted' to a locally conformal-invariant (Weyl-invariant) extension, by introducing an additional singlet scalar field along with local Weyl symmetry. Restricting consideration to the Einstein-Higgs part of the action, and for the moment ignoring the gauge fields, the usual Einstein-Higgs minimal coupling action,

$$ \mathcal{S} = \int \sqrt{-g} \left[\frac{M_{Pl}^2}{2} R - \frac{1}{2} (\partial h)^2 - V(h) \right], \tag{6} $$

can be viewed as a gauge fixed version of following action:

$$ \mathcal{S} = \int \sqrt{-g} \left[\frac{1}{12} (\phi^2 - H^2) R + \frac{1}{2} \left((\partial \phi)^2 - (\partial H)^2 \right) - V(H/\phi) \right], \tag{7} $$

which is invariant under classical Weyl transformations, $\phi(x) \rightarrow \Omega^{-1}(x)\phi(x)$, $H(x) \rightarrow \Omega^{-1}(x)H(x)$, $g_{\mu\nu} \rightarrow \Omega^2(x)g_{\mu\nu}(x)$. For, we are free to choose Einstein gauge $\phi^2 - H^2 = 6M_{Pl}^2$, which implies that there is a field h satisfying $\Phi = \sqrt{6M_{Pl}} \cosh(h/\sqrt{6M_{Pl}})$ and $H = \sqrt{6M_{Pl}} \sinh(h/\sqrt{6M_{Pl}})$. Making these substitutions then, at least for $H \ll \phi$, we recover the original action (6).

It will be noticed that the kinetic term for ϕ in (7) has the wrong sign: ϕ is a 'ghost' field. However, if Weyl symmetry holds there is no problem since we can always pick the Weyl gauge $\phi = M_{Pl}$ to remove any associated instability. Even more interesting, if we are studying FRW cosmologies we can, through a Weyl gauge transformation, choose 'unimodular' gauge, in which the determinant of the metric is set constant (or the scale factor a is set equal to unity), so that the metric becomes completely nonsingular all the way to the FRW singularity. By considering the gauge-invariant quantity $(\phi^2 - H^2)(-g)^{1/4}$, one sees that the Einstein-gauge scale factor a_E and the values of the two scalar fields in unimodular gauge, ϕ_γ and H_γ, are related via $a_E^2 = \phi_\gamma^2 - H_\gamma^2$, when this quantity is positive. The space (ϕ_γ, H_γ) has a two-dimensional Minkowski metric as is seen from the kinetic terms in (7). On this space, lines of constant a_E are hyperbolae, with the FRW singularities $a_E = 0$ separating the regions with $\phi^2 > H^2$ from the regions with $\phi^2 < H^2$ (Fig. 14). The former regions have 'normal gravity,' in the sense that the coefficient of the Einstein term in the action is positive, whereas the latter regions have 'antigravity,' because the coefficient of the Einstein term in the action is negative. In the latter regions, the

Einstein-gauge scale factor is given by $a_E^2 = H_\gamma^2 - \phi_\gamma^2$. When Newton's constant becomes negative, one should worry about the stability of the vacuum, since if one treats the antigravity region locally as a patch of flat space, gravitons will have negative kinetic energy and can be pair created along with, say, positive energy photons. However, the antigravity region is merely the classical low-energy description of a short-lived intermediate state. Calculations show that when the asymptotic states of the graviton are defined in the gravity region, there is no such instability.[18]

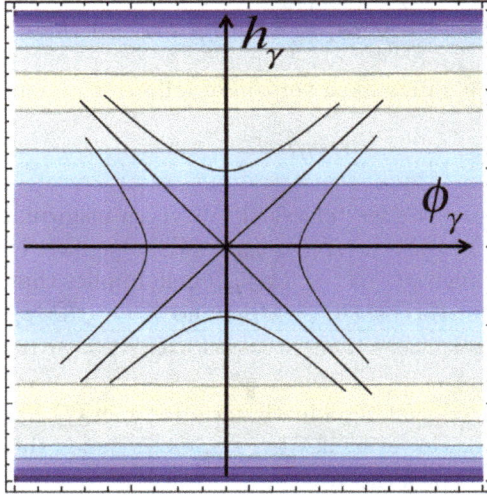

Fig. 14. The space of dynamical fields in unimodular gauge (indicated by the subscript γ). The light cone corresponds to vanishing Einstein frame scale factor. Superposed on the plot is a contour plot of the electroweak Higgs potential, as approximated in (8).

We have extensively studied the complete set of classical FRW solutions in such theories, showing that within unimodular gauge they are generically regular and geodesically complete.[19,20] Recently, we have extended these solutions to potentials of the same form as the electroweak Higgs potential, i.e. with a local minimum separated by a barrier from a region of instability. A simplified action for the Einstein-Higgs evolution of a flat FRW universe is:

$$\mathcal{S} = \int d\tau \left[\frac{1}{2e}(\partial_\tau(ah))^2 - \frac{1}{2e}(\partial_\tau(a\phi))^2 - e(\lambda(ah)^4 - \rho_r - \sqrt{\rho_r}(ah)^2 - \Lambda a^4(\phi^2 - h^2)^2) \right], \tag{8}$$

where e is the 'lapse' function associated with time reparameterisation invariance, whose variation yields the Friedmann equation. This action is invariant under conformal transformations $a \to \Omega(\tau)a$, $\phi \to \Omega(\tau)\phi$, $h \to \Omega(\tau)h$. The terms in the potential are as follows. The term λh^4, with λ negative, represents the Higgs potential for h of order unity, as illustrated in Fig. 12. For simplicity, I shall ignore the terms describing the symmetry breaking minimum and the barrier at $h \sim 10^{-6}$. The constant ρ_r represents the density of radiation — in Einstein gauge the radiation density is ρ_r/a^4. The term involving $\sqrt{\rho_r}h^2$ represents the temperature-dependent mass term of the Higgs field due to its coupling to radiation. The final term, with Λ a small, negative dimensionless constant, represents the effect of a negative Einstein-frame cosmological constant. It is included in order to crudely represent the effect of the decay of today's dark energy, either through an ekpyrotic phase (driven by another scalar field) or by the nucleation of a bubble representing the decay of today's Higgs vacuum to a lower energy state.

Figure 15 shows the evolution of a generic initial configuration, in unimodular gauge. The universe is radiation dominated at early times. In the antigravity region, the Einstein-frame scale factor is given by $a_E^2 = h_\gamma^2 - \phi_\gamma^2$. The sign reversal has a key consequence: that the presence of radiation speeds the passage through the 'antigravity' region. This may be seen from the Friedmann equation in Einstein frame, which reads $a_E'^2 = a_E^2 h'^2/6 + \rho_r/3 + \ldots$ in the gravity region, so a larger ρ_r leads to a larger expansion rate there, but reads $a_E^2 h'^2/6 = a_E'^2 + \rho_r + \ldots$ in the gravity region, so a larger ρ_r leads to a larger velocity of the Einstein-frame scalar h, $i.e.$ along the $a_E =$ constant hyperbolae in the antigravity region. In this way, a radiation-dominated universe spends only of order one dynamical (Hubble) time in the antigravity region, before emerging smoothly into the gravity region. As long as the Higgs field starts out small enough, the temperature-dependent mass stabilises it around small values, and it oscillates stably as the universe expands. When the Einstein-frame scale factor becomes large, the negative cosmological constant dominates and causes the universe to recollapse to a crunch, pass through the antigravity phase once more and re-emerge in a second big bang. The cycles continue stably for an apparently unlimited time, gradually filling out a torus in phase space. The lower plot in Fig. 16 shows a cross-section of the torus, in the $(\phi_\gamma, \dot{\phi}_\gamma)$ plane, exhibiting classic Kolmogorov-Arnold-Moser behaviour. That is, the cyclic solution is quasiperiodic, rather than chaotic, due to the incommensurate periods of the Higgs oscillations and the oscillations of the

24

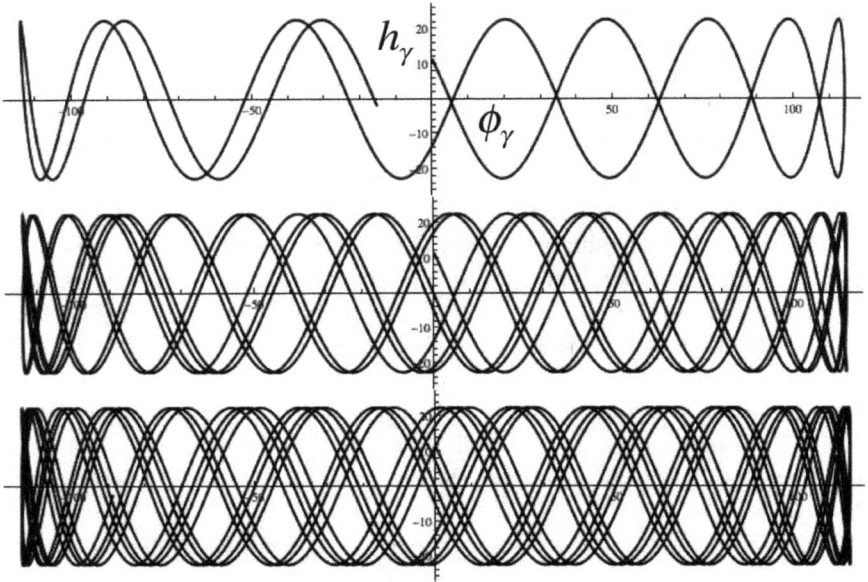

Fig. 15. The evolution of an FRW universe containing a Higgs field with radiation and cosmological constant, as described in unimodular gauge. The evolution is completely regular and cyclic. The sequence continues in Fig. 16.

scale factor under the combined influence of the radiation and cosmological constant.

These plots show the evolution of an adiabatic bouncing universe, in the presence of a classically metastable Higgs field. They demonstrate that a Higgs potential, of the type implied by the LHC measurements, is perfectly compatible with cyclic evolution of the universe. Provided the Higgs field value lies within a 'stability band', around $h = 0$, it will oscillate stably as the universe undergoes many cycles of expansion and contraction.

For a realistic cyclic cosmology, we have to do more. In particular, we have to describe the evolution of a generic, perturbed universe through crunch/bang transitions and we have to describe non-adiabatic, entropy generating processes. Regarding the former, it is notable that the presence of the Higgs field is enough to remove chaotic Mixmaster behaviour of the type exhibited by pure Einstein gravity coupled to radiation. The approach to cosmic singularities is generically smooth and Kasner-like, exhibiting 'ultralocal' evolution as an anisotropic Bianchi-type universe. We have studied this problem in detail, finding that although the evolution

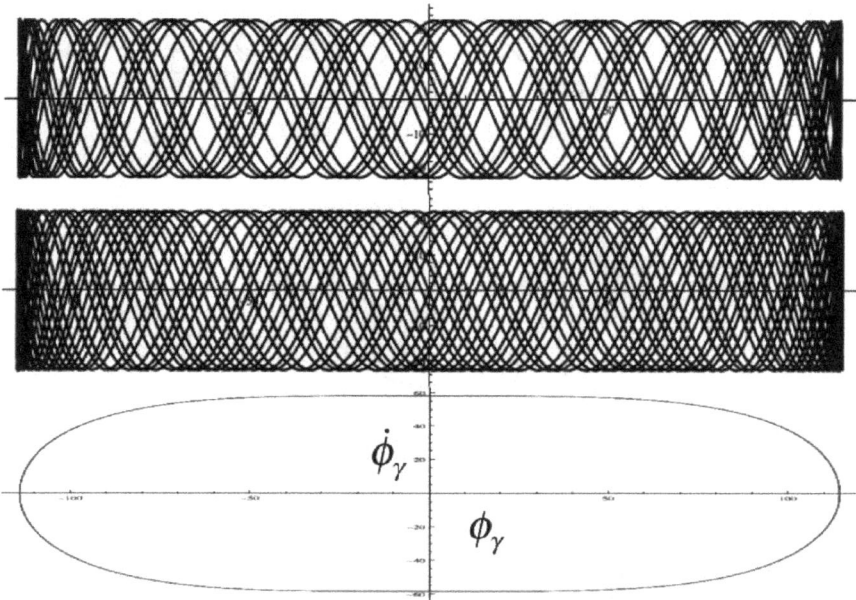

Fig. 16. Continued evolution of the FRW-Higgs system in unimodular gauge. The last plot shows the projection of the trajectory onto the $(\phi_\gamma, \dot{\phi}_\gamma)$ plane, clearly exhibiting the localisation of the trajectory on a torus within phase space, as expected from the KAM theorem.

across the light cone in the (ϕ_γ, h_γ) plane is no longer analytic, it is still unique.[19] In principle, this solves the problem of the classical evolution of cosmological perturbations within such a cyclic universe, but further work is needed to study the detailed consequences.

More work is also needed on non-adiabatic processes such as the production of radiation around the crunch-bang transition. We have crudely modelled this by increasing the parameter ρ_r and matching the scale factor and its first derivative, at some arbitrary point in the antigravity region, in every cycle. The resulting solutions describe a universe whose size grows with each cycle, reaching larger and larger values of the Einstein-frame scale factor a_E.

6. Holographic Description of a Bouncing Cosmology

The above description of a bouncing cosmology has the significant drawback that it is strictly given within the framework of effective field theory. In

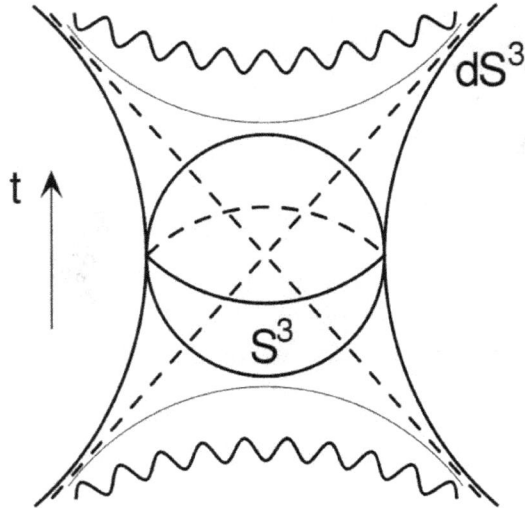

Fig. 17. A cosmology consisting of an infinite, open FRW bubble nucleating within an asymptotically-anti-de Sitter (AdS^4) bulk. The boundary is taken to be a three-dimensional de Sitter (dS^3) spacetime, which avoids the big bang and big crunch singularities. A dual theory on this boundary provides, in principle, a complete holographic description.[23]

particular, one may question whether higher derivative corrections become crucial around the big crunch or big bang, and whether the 'antigravity' phase is real or just an artefact of the low energy approximation. To address these questions, it is desirable to study a big crunch/big bang transition in a framework where stringy and quantum gravity effects may be taken into account, within a theory which has some claim to being ultraviolet-complete. In recent work, Michael Smolkin and I have studied the description of a big crunch cosmology within M-theory, using the AdS/CFT correspondence.

The setup is illustrated in Fig. 17. We consider an M-theory bulk, in which eleven-dimensional supergravity is compactified on a seven-dimensional sphere. The seven sphere is represented as a circle — the M-theory dimension — fibered over CP^3. In the four-dimensional low-energy description, there is a scalar modulus parameterising the radius of the M-theory circle. This field has a tachyonic mass and a potential which is unbounded below. If supersymmetric boundary conditions are imposed on the bulk, the scalar field is stabilised and the ground state for the bulk is four-dimensional anti-de Sitter spacetime. But if more general, non-supersymmetric but still AdS-invariant boundary conditions are imposed, cosmological evolution is allowed in bulk. In particular, the scalar field is free to run downhill within a cosmological bubble whose interior is an infinite, open, FRW cosmology. The complete asymptotically-AdS bulk contains both a big bang FRW cosmology to the past and a big crunch FRW cosmology to the future.

The AdS-dual to this cosmology consists of a deformation of a three-dimensional supersymmetric Chern-Simons gauge theory known as ABJM theory.[21] In the limit where the gauge coupling is small, the AdS bulk curvature scale is smaller than the string scale. The bulk is very difficult to describe in this regime, but the boundary theory is weakly coupled and analytically tractable. In fact, it reduces to the well-known three-dimensional $O(N)$-invariant vector model. This theory possesses a nontrivial ultraviolet fixed point, suggesting it is capable of providing a complete description of the bulk dynamics.[22] A longstanding puzzle, however, has been that in flat space the theory seems to be ill-defined, because due to quantum effects the effective potential is unbounded below. Our analysis confirms that the theory is ill-defined when quantised on flat space. However, on three dimensional de Sitter (dS^3) spacetime, we have been able to make sense of the theory by carefully defining the path integral on a suitable complex contour in field space. At the fixed point coupling, this generalised boundary field theory exhibits exact Weyl invariance which is spontaneously broken. One can compute the field correlation functions, and continue them to the future conformal boundary of the dS^3, which is a two-sphere. Just as the AdS-invariance of the bulk implies conformal invariance of the three-dimensional boundary theory, so de Sitter invariance of the boundary dS^3 implies two-dimensional conformal symmetry on *its* two boundary two-spheres. One can glue another copy of dS^3 onto the first by identifying the future conformal boundary of the first dS^3 with the past conformal boundary of the second. Conformal invariance then imposes a unique matching rule identifying the outgoing data on the first dS^3 with the incoming data on the second dS^3. In this way, conformal symmetry yields a unique evolution rule for passing 'around' the big crunch singularity within the bubble in the interior of the asymptotically-AdS spacetime. It will be fascinating to see whether this matching rule agrees with the rule found from studies of the low-energy effective theory, as presented above. The holographic description involves no intermediate 'antigravity' phase, but this phase may nevertheless be present from the point of view of the low-energy effective theory description of the bulk.

7. Summary and Conclusions

The idea of spontaneous symmetry breaking has to count as one of the most remarkable human intellectual achievements. The discovery of the Higgs boson confirms these deep insights into nature. At the time of writing, the Nobel prize in physics has just been awarded, to Francois Englert and Peter Higgs. While celebrating this richly-deserved recognition, we also celebrate the foundational contributions made by Tom Kibble. His elegant and farsighted work anticipated the profound consequences of spontaneous symmetry breaking and phase transitions in cosmology. Inflationary theory emerged from these ideas, and has been hugely successful as a phenomenological model of the hot big bang and the formation of cosmic structure. Nevertheless, the story is far from complete. Not only gauge theory, but also perhaps conformal symmetry, may have been spontaneously broken in nature. And if that is the case, we can look forward to a complete and consistent picture not only of the fundamental physical laws, but of the cyclic evolution of the universe. A very happy birthday, Tom!

References

1. T. W. B. Kibble, *Scholarpedia* **4**(1):6441 (2009).
2. T. W. B. Kibble, *J. Phys. A* **9**, 1387 (1976).
3. T. W. B. Kibble, *Phys. Rep.* **67**, 183 (1980).
4. For a summary, see K. Rummukainen, K. Kajantie, M. Laine, M. E. Shaposhnikov and M. Tsypin, hep-ph/9809435.
5. A. H. Guth, *Phys. Rev. D* **23**, 347 (1981).
6. A. D. Linde, *Phys. Lett. B* **114**, 431 (1982).
7. A. Albrecht and P. J. Steinhardt, *Phys. Rev. Lett.* **48**, 1220 (1982).
8. D. Larson *et al.*, *Astrophys. J. Suppl.* **192**, 16 (2011) [arXiv:1001.4635]; E. Komatsu *et al.* [WMAP Collaboration], *Astrophys. J. Suppl.* **192**, 18 (2011) [arXiv:1001.4538].
9. D. Coulson, R. G. Crittenden and N. G. Turok, *Phys. Rev. Lett.* **73**, 2390 (1994) [astro-ph/9406046].
10. G. W. Gibbons and N. Turok, *Phys. Rev. D* **77**, 063516 (2008) [hep-th/0609095].
11. A. A. Starobinsky, *Phys. Lett. B* **91**, 99 (1980).
12. P. J. Steinhardt, *Scientific American*, April 2011, p. 36.
13. U.-L. Pen, U. Seljak and N. Turok, *Phys. Rev. Lett.* **79**, 1611 (1997) [astro-ph/9704165].
14. For a recent summary, see J.-L. Lehners and P.J. Steinhardt, arXiv:1304.3122.
15. I. Bars, P. J. Steinhardt and N. Turok, arXiv:1307.8106 [gr-qc].
16. J. Khoury, P. J. Steinhardt and N. Turok, *Phys. Rev. Lett.* **91**, 161301 (2003).
17. D. Buttazzo, G. Degrassi, P. P. Giardino, G. F. Giudice, F. Sala, A. Salvio and A. Strumia, arXiv:1307.3536 [hep-ph].

18. N. Turok, unpublished.

19. I. Bars, S.-H. Chen, P. J. Steinhardt and N. Turok, *Phys. Lett. B* **715**, 278 (2012) [arXiv:1112.2470 [hep-th]].

20. I. Bars, S.-H. Chen, P. J. Steinhardt and N. Turok, *Phys. Rev. D* **86**, 083542 (2012) [arXiv:1207.1940 [hep-th]].

21. O. Aharony, O. Bergman, D. L. Jafferis and J. Maldacena, *JHEP* **0810**, 091 (2008) [arXiv:0806.1218 [hep-th]].

22. B. Craps, T. Hertog and N. Turok, *Phys. Rev. D* **80**, 086007 (2009) [arXiv:0905.0709 [hep-th]].

23. M. Smolkin and N. Turok, arXiv:1211.1322 [hep-th]; in preparation (2013).

UNIVERSALITY OF PHASE TRANSITION DYNAMICS: TOPOLOGICAL DEFECTS FROM SYMMETRY BREAKING

ADOLFO DEL CAMPO[*,†] and WOJCIECH H. ZUREK[*]

*Theoretical Division, Los Alamos National Laboratory, USA
†Center for Nonlinear Studies, Los Alamos National Laboratory, USA

In the course of a nonequilibrium continuous phase transition, the dynamics ceases to be adiabatic in the vicinity of the critical point as a result of the critical slowing down (the divergence of the relaxation time in the neighborhood of the critical point). This enforces a local choice of the broken symmetry and can lead to the formation of topological defects. The Kibble–Zurek mechanism (KZM) was developed to describe the associated nonequilibrium dynamics and to estimate the density of defects as a function of the quench rate through the transition. During recent years, several new experiments investigated the formation of defects in phase transitions induced by a quench both in classical and quantum mechanical systems. At the same time, some established results were called into question. We review and analyze the Kibble–Zurek mechanism focusing in particular on this surge of activity, and suggest possible directions for further progress.

1. Introduction

The aim of this paper is to provide a limited review of the experiments that test the Kibble–Zurek mechanism (KZM): we shall focus on the experiments that test the scaling of the number of topological defects with the quench rate predicted by the KZM. This self-imposed restriction limits the number of the relevant experiments to a manageable total. It is also a sign that the field — that has its roots in the seminal papers of Tom Kibble[1,2] — has matured, so that the question that was initially most pressing (i.e. whether topological defects form at all via KZM) has been by now answered in the affirmative in a variety of systems,[3–16] although ^{4}He remains a confounding exception.[17,18]

The scaling of the defect density with the quench rate — prediction of the *nonequilibrium* effect using *equilibrium* critical exponents[19,20] — is the key testable consequence of the KZM. However, the resulting dependence of the size of the domains where symmetry can be broken "in unison" is usually given by a power law with a small fractional exponent. Therefore, to detect a significant variation in the defect density one needs to vary quench rates over a large range. This tends to be difficult in the traditional thermodynamic phase transition experiments. For instance, cooling (that can lead to a symmetry breaking transition) will typically result in temperature gradients inside the bulk of the system that can suppress defect formation,[21,22] but it can also drive convection that can create defects, such as vortex lines in superfluids, independently of the KZM.[17,18]

There are several reviews of the subject starting with[23] and more recent monographs[24–28] that discuss the KZM, its consequences, and related phenomena in phase transitions. As is also the case with this review, all of these reviews cover only selected fragments of the field either because (as a result of recent developments) they are out of date, or because they are focused on specific subfields (e.g. quantum phase transitions). We focus on the (mostly recent) experiments that test scalings predicted by the KZM and the related theoretical developments.

2. The Kibble–Zurek Mechanism

Consider the dynamics of spontaneous symmetry breaking in the course of a phase transition induced by the change of a control parameter λ. A continuous second-order phase transition is characterized by the divergence (usually as a power-law) of both the *equilibrium* correlation length ξ

$$\xi(\varepsilon) = \frac{\xi_0}{|\varepsilon|^\nu}, \tag{1}$$

and *equilibrium* relaxation time τ

$$\tau(\varepsilon) = \frac{\tau_0}{|\varepsilon|^{z\nu}}, \tag{2}$$

as a function of the distance to the critical point λ_c. It is convenient to define the reduced distance parameter

$$\varepsilon = \frac{\lambda_c - \lambda}{\lambda_c}, \tag{3}$$

in terms of which the system initially prepared in the high-symmetry phase ($\varepsilon < 0$) is forced to face a spontaneous symmetry breaking scenario as the critical point is crossed towards the degenerate vacuum manifold ($\varepsilon > 0$).

In Eq. (1), ν is the correlation length critical exponent, while z in Eq. (2) is the dynamic critical exponent. Different systems belonging to the same universality class share the same critical exponents. Above, ξ_0 and τ_0 are dimensionful constants that depend on the microphysics in contrast with ν and z that depend only on the universality class of the transition. The Kibble–Zurek mechanism (KZM) describes the dynamics of a continuous phase transition under a time-dependent change of λ across the critical value. The time-dependence $\lambda(t)$ in the proximity of λ_c can usually be linearized. Therefore, we assume a linear quench

$$\lambda(t) = \lambda_c[1 - \varepsilon(t)] \tag{4}$$

symmetric around the critical point so that the reduced parameter is characterized by the quench time τ_Q and varies linearly in time according to

$$\varepsilon(t) = \frac{t}{\tau_Q}, \tag{5}$$

in $t \in [-\tau_Q, \tau_Q]$, the critical point being reached at $t = 0$. Far away from the critical point $|\lambda| \gg \lambda_c$, the equilibrium relaxation time is very small with respect to the time remaining until reaching the critical point following the quench (5), and the dynamics is essentially adiabatic. In the opposite limit, in the close neighbourhood of $\varepsilon(t) = 0$, the dynamics is approximately frozen due to the divergence of the equilibrium relaxation time (critical slowing down). The system is then unable to adjust to the externally imposed change of the reduced control parameter $\varepsilon(t)$. Exploiting this intuition,[19] the KZM splits the dynamics into the sequence of three stages where the dynamics is adiabatic, effectively frozen, and adiabatic again, as $\varepsilon(t)$ is varied from $\varepsilon(t) < 0$ to $\varepsilon(t) > 0$. See Fig. 1 for a schematic representation.

This simplification, often referred to as the adiabatic-impulse approximation, captures the essence of the nonequilibrium dynamics involved in the crossing of the phase transition at a finite rate. The inability of the collective degree of freedom that defines the order parameter to keep up with the change imposed from the outside is the essence of the freeze-out. This does not mean that all of the evolution in the system stops, or even that the evolution of the order parameter ceases completely: the microstate of the system will of course evolve as dictated by its (time-dependent) Hamiltonian, and even the local thermodynamic equilibrium of the microscopic

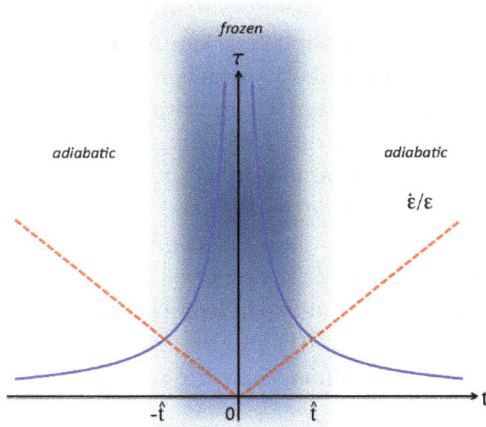

Fig. 1. Schematic representation of the freeze-out captured by the adiabatic-impulse approximation. During a linear quench, the reduced control parameter $\varepsilon = t/\tau_Q$ forces the system to cross the critical point from the high symmetry phase ($t < 0$) to the low symmetry phase ($t > 0$). Due to divergence of the equilibrium relaxation time, associated with the critical slowing down in the neighbourhood of $\varepsilon = 0$, the order parameter of the system ceases to follow the equilibrium expectation value and enters an impulse stage within the time interval $[-\hat{t}, \hat{t}]$.

degrees of freedom may be maintained. However, the order parameter will cease to follow its equilibrium value, and it will be able to catch up with it locally, to the extent allowed by the presence of topological defects, only after the critical point has been passed, usually with a delay of about $\hat{t}$, as illustrated, for example, by numerical simulations of BEC formation.[29]

The boundary between the adiabatic and frozen stages can be estimated by comparing the equilibrium relaxation time with the time elapsed after crossing the critical point

$$\tau(t) \approx |\varepsilon/\dot{\varepsilon}| = t \,. \tag{6}$$

This equation[19] yields the time scale

$$\hat{t} \sim \left(\tau_0 \tau_Q^{z\nu}\right)^{\frac{1}{1+z\nu}} \,, \tag{7}$$

known as the freeze-out time. The degrees of freedom of the system relevant for the selection of broken symmetry cannot keep up with the externally imposed change of ε, and, consequently, the order parameter of the system lags behind its equilibrium value corresponding to the instantaneous value of ε within the interval $\varepsilon \in [-\hat{\varepsilon}, \hat{\varepsilon}]$, where

$$\hat{\varepsilon} \equiv |\varepsilon(\hat{t})| \sim \left(\frac{\tau_0}{\tau_Q}\right)^{\frac{1}{1+z\nu}} . \tag{8}$$

Spontaneous symmetry breaking entails degeneracy of the ground state. In an extended system, causally disconnected regions will make independent choices of the vacuum in the new phase. A summary of the topological classification of the resulting defects using homotopy theory is presented in Appendix A. The KZM sets the average size of these domains by the value of the equilibrium correlation length at $\hat{\varepsilon}$,[19]

$$\hat{\xi} \equiv \xi[\hat{\varepsilon}] = \xi_0 \left(\frac{\tau_Q}{\tau_0}\right)^{\frac{\nu}{1+z\nu}} . \tag{9}$$

This is the main prediction of the KZM.

This simple form of a power law of $\hat{t}$ (and, consequently, of $\hat{\xi}$) arises only when the relaxation time of the system scales as a power law of ε. This need not always be the case. For example, in the Kosterlitz-Thouless phase transition universality class, of relevance to 2D Bose gases, the critical slowing down is described by a more complicated (exponential) dependence on ε. A more complex dependence of $\hat{t}$ and $\hat{\xi}$ on τ_Q (rather than a simple power law) would be then predicted as a result.[30]

The above estimate of the $\hat{\xi}$ is often recast as an estimate for the resulting density of topological defects,

$$n \sim \frac{\hat{\xi}^d}{\hat{\xi}^D} = \frac{1}{\xi_0^{D-d}} \left(\frac{\tau_0}{\tau_Q}\right)^{(D-d)\frac{\nu}{1+z\nu}} , \tag{10}$$

where D and d are the dimensions of the space and of the defects (e.g. $D = 3$ and $d = 1$ for vortex lines in a 3D superfluid). This order-of-magnitude prediction usually overestimates the real density of defects observed in numerics. A better estimate is obtained by using a factor f, to multiply $\hat{\xi}$ in the above equations, where $f \approx 5 - 10$ depends on the specific model.[29,31–35] Thus, while KZM provides an order-of-magnitude estimate of the density of defects, it does not provide a precise prediction of their number. However, if one were able to check the power law above, one could claim that the KZM holds and show that the nonequilibrium dynamics across the phase transition is also universal. This requires the ability to measure the average number of excitations after driving the system at a given quench rate, and repeating this measurement for different quench rates.

3. Landau–Zener Crossing as a Quantum Example of the KZM

Landau[36] and Zener[37] (see as well Stueckelberg[38] and Majorana[39]) provided an analytical description of the diabatic excitation probability in a two-level quantum mechanical system described by a Hamiltonian $\hat{\mathcal{H}}_0$ in which the energy gap between the two states varies linearly in time. Using dimensionless units for all variables,

$$\hat{\mathcal{H}}_0 = \begin{pmatrix} t/\tau_Q & 1 \\ 1 & -t/\tau_Q(t) \end{pmatrix} = \frac{t}{\tau_Q}\sigma^z + \sigma^x \,, \tag{11}$$

where $\sigma^{x,y,z}$ are the usual Pauli matrices, and for which the instantaneous eigenbasis reads:

$$|\uparrow(t)\rangle = \sin(\theta/2)|1\rangle + \sin(\theta/2)|2\rangle,$$
$$|\downarrow(t)\rangle = -\sin(\theta/2)|1\rangle + \cos(\theta/2)|2\rangle \,.$$

The angle $\theta \in [0, \pi]$ obeys the relations

$$\cos\theta = \frac{\varepsilon}{\sqrt{1+\varepsilon^2}}, \quad \sin\theta = \frac{1}{\sqrt{1+\varepsilon^2}} \,,$$

in terms of the reduced variable

$$\varepsilon = \frac{t}{\tau_Q} \,. \tag{12}$$

The exact energy gap is $E_\uparrow(t) - E_\downarrow(t) = \sqrt{1+\varepsilon^2}$. The Landau–Zener (LZ) formula states that the excitation probability decays exponentially with the quench time

$$P = e^{-\frac{\pi}{2}\tau_Q} \,. \tag{13}$$

Above, time is measured in units given by the inverse of the gap in Eq. (11) at its minimum. This result has been extended to multi-state problems[40–43] as well as nonlinear modulations of $\varepsilon(t)$.[44,45]

Damski has shown that the quantum dynamics across a Landau–Zener (LZ) transition is accurately described by the adiabatic-impulse approximation, and ultimately, by the KZM.[46] The freeze-out time scale can be estimated by matching the inverse of the energy gap with the time scale $|\varepsilon/\dot{\varepsilon}|$

$$\frac{1}{\sqrt{1+(\hat{t}/\tau_Q)^2}} = \alpha\hat{t} \tag{14}$$

where α is a constant. It follows that $\hat{\varepsilon} = \hat{t}/\tau_Q = \frac{1}{\sqrt{2}}\left[\sqrt{1+\left(\frac{1}{\alpha\tau_Q}\right)^2}-1\right]^{1/2}$.
One can then consider the case where the system is initialized at a time $t_i \ll -\hat{t}$ and evolved until a final time $t_f \gg \hat{t}$. The impulse stage occurs in the interval $[-\hat{t}, \hat{t}]$ and the excitation probability can then be approximated by

$$P = |\langle\uparrow(\hat{t})|\downarrow(-\hat{t})\rangle|^2 = \frac{\hat{\varepsilon}^2}{1+\hat{\varepsilon}^2}. \tag{15}$$

Using the estimate for $\hat{\varepsilon}$, one finds that $P = 1 - \alpha\tau_Q/2 + (\alpha\tau_Q)^2/2 + \dots$. The optimal value $\alpha = \pi/2$ can be extracted from the comparison with the exact solution of the LZ problem.[47] This result agrees with the LZ formula up to third order in τ_Q.

Exploiting the adiabatic impulse approximation, one can consider as well asymmetric quenches, such as when $t_i = 0$, for which

$$P = |\langle\uparrow(\hat{t})|\downarrow(0)\rangle|^2 = \frac{1}{2} - \frac{1}{2\sqrt{1+\hat{\varepsilon}^2}}. \tag{16}$$

Its expansion, $P = \frac{1}{2} - \frac{1}{2}\sqrt{\alpha\tau_Q} + \frac{1}{8}(\alpha\tau_Q)^{3/2} + \dots$, matches well the exact result for $\alpha \simeq \pi/4$.[47]

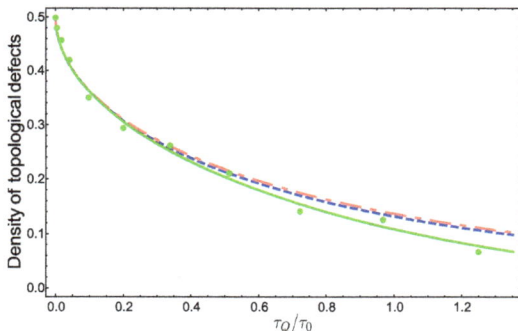

Fig. 2. Experimental optical simulation of the quantum dynamics across a LZ crossing supporting the adiabatic-impulse approximation. The measured density of excitations (green dots) agrees with the exact solution (solid green line) and the estimate based on the adiabatic-impulse approximation.[46] From Xu *et al.*[48]

An experimental demonstration of the KZM–LZ connection,[46] the possibility of describing a LZ crossing using the adiabatic-impulse approximation which is a core feature of the KZM, has recently been achieved using a linear optical quantum simulator at the Key Laboratory of Quantum Information.[48] A second experiment in the same center, this time in a semiconductor electron charge qubit, has further confirmed the universal validity of the adiabatic-impulse approximation.[49]

3.1. *Controlling excitations in Landau–Zener crossing*

Excitations formed during a LZ crossing at an arbitrary finite-rate can be completely suppressed by counterdiabatic driving. This method was introduced by Demirplak and Rice,[50] and Berry,[51] and is also referred to as the transitionless quantum driving. Provided that one can diagonalize the Hamiltonian of interest $\hat{\mathcal{H}}_0[\lambda(t)]$ (that is, find its instantaneous eigenstates $|n(\lambda)\rangle$ and eigenvalues $E_n(\lambda)$) for every $\lambda(t)$, it is possible to enforce the dynamics exactly through the adiabatic manifold using counterdiabatic fields (i.e. the fields that allow one to cross the adiabatic-impulse regime fast, but without the usually inevitable excitations). Indeed, the adiabatic approximation

$$\psi_n(t) = \exp\left(-\frac{i}{\hbar}\int_0^t E_n(t')dt' - \frac{1}{\hbar}\int_0^t \langle n|\partial_{t'} n\rangle dt'\right)|n(t)\rangle \qquad (17)$$

to $\hat{\mathcal{H}}_0[\lambda(t)]$ becomes the exact solution of the time-dependent Schrödinger equation with the Hamiltonian $\hat{\mathcal{H}} = \hat{\mathcal{H}}_0 + \hat{\mathcal{H}}_1$, where

$$\hat{\mathcal{H}}_1 = i\lambda'(t)\sum_n [|\partial_\lambda n\rangle\langle n| - \langle n|\partial_\lambda n\rangle|n\rangle\langle n|]. \qquad (18)$$

Counterdiabatic driving has been demonstrated experimentally in an effective two-level system realized with a Bose–Einstein condensate in the presence of an optical lattice potential.[52] This type of assisted quantum adiabatic passage has also been implemented in an electron spin of a single nitrogen-vacancy center in diamond.[53] For the LZ crossing with $\lambda(t) = t/\tau_Q$, one finds that the counterdiabatic field reduces to

$$\hat{\mathcal{H}}_1 = -\frac{1}{2\tau_Q}\frac{\Delta}{1 + (t/\tau_Q)^2}\sigma^y. \qquad (19)$$

Counterdiabatic driving is currently finding an increasing number of applications in quantum control,[54] quantum information processing,[53] BEC and ultra cold atom physics,[55] and other fields.[56]

4. Quantum Phase Transitions

We have seen that two-level systems constitute an ideal platform to test the adiabatic-impulse approximation, a key ingredient of the quantum KZM.

However, KZM also predicts the typical size of the domains in the broken symmetry phase resulting from a finite-rate crossing of a critical point, i.e. it estimates the average distance between topological defects. To analyze this aspect it is required to consider spatially extended systems. A wide variety of condensed-matter systems and statistical mechanics models exhibiting quantum phase transitions offer a test-bed for these predictions.

Quantum phase transitions are characterized by abrupt changes in the ground-state properties of many-body systems as a control parameter is tuned.[57] In experimental realizations this control parameter is typically an external field such as a magnetic field acting on spins, a laser field in trapped ion systems or and optical lattice potential in ultracold atom quantum simulators.[57,58]

The extension of the KZM to quantum phase transitions was elucidated by studying the dynamics in quasi-free fermion models,[59–62] and is by now well-documented.[27,28] A paradigmatic example is the one-dimensional quantum Ising chain described by the Hamiltonian

$$\hat{\mathcal{H}} = -\sum_{k=1}^{N} \left[g(t)\sigma_n^x + \sigma_n^z \sigma_{n+1}^z \right], \tag{20}$$

where $g(t)$ plays the role of a magnetic field, which has a critical point at $|g_c| = 1$. Remarkably, this model describes certain magnetic condensed matter systems[63] and its quantum emulation, e.g. in ion traps, is the subject of ongoing efforts.[64] A quantum phase transition occurs between a paramagnetic phase ($|g| > 1$) and a doubly-degenerate ferromagnetic phase ($|g| < 1$).

Consider the time-dependent quench $g(t) = -t/\tau_Q$ with $t \in (-\infty, 0)$. One can quantify the breakdown of adiabaticity dictated by the KZM using the average number of excitations for a given quench rate ending at $g = 0$,

$$n = \frac{1}{2N} \sum_{k=1}^{N} [1 - \langle \sigma_n^z \sigma_{n+1}^z \rangle]. \tag{21}$$

Using standard techniques (a combination of the Jordan–Wigner transformation and Fourier transform), Dziarmaga was able to rewrite the system as a set of independent Landau–Zener crossings.[60] In the thermodynamic limit ($N \gg 1$), the density of kinks can then be approximated by

$$n = \frac{1}{2\pi} \int_{-\pi}^{\pi} p_k dk, \tag{22}$$

where p_k is the probability of excitation in each mode. In view of the applicability of the adiabatic-impulse approximation to each level, the dynamics across the critical point might be expected to be described by the KZM. The resulting amount of excitations is found to scale as $n \propto \tau_Q^{-1/2}$. This result is based on an exact solution of the dynamics for the Ising model.[47,60] However, it can be extended to an arbitrary D dimensional Hamiltonian $\hat{\mathcal{H}}[\lambda(t)]$, with a quantum critical point characterized by critical exponents ν and z, leading to the estimate $n \propto \tau_Q^{-\frac{(D-d)\nu}{\nu z+1}}$, see Ref. 61 and the reviews.[27,28]

5. Adiabatic Crossing of Quantum Phase Transition

Counterdiabatic driving[50,51] has been extended to many-body systems and to quasi-free fermion models exhibiting a quantum phase transition.[65] Consider the family of D dimensional model Hamiltonians, which can be decomposed into the sum of uncoupled $\mathbf{k}$-mode Hamiltonians,

$$\hat{\mathcal{H}}_0 = \sum_{\mathbf{k}} \psi_{\mathbf{k}}^{\dagger} \left[\vec{a}_{\mathbf{k}}(\lambda(t)) \cdot \vec{\sigma}_{\mathbf{k}}\right] \psi_{\mathbf{k}}, \tag{23}$$

where the Pauli matrices in the mode $\mathbf{k}$ are $\vec{\sigma}_{\mathbf{k}} \equiv (\sigma_{\mathbf{k}}^x, \sigma_{\mathbf{k}}^y, \sigma_{\mathbf{k}}^z)$. $\psi_{\mathbf{k}}^{\dagger} = (c_{\mathbf{k},1}^{\dagger}, c_{\mathbf{k},2}^{\dagger})$ are fermionic operators. The function $\vec{a}_{\mathbf{k}}(\lambda) \equiv (a_{\mathbf{k}}^x(\lambda), a_{\mathbf{k}}^y(\lambda), a_{\mathbf{k}}^z(\lambda))$ is specific for each model.[27] Examples of quantum critical models within this family are the Ising and XY models[57] in $D = 1$, and the Kitaev model in $D = 1, 2$.[66,67] As quasi-free fermion models, they can be written down as a sum of independent Landau–Zener crossings. The dynamics across the quantum critical point can be driven through the adiabatic solution associated with $\hat{\mathcal{H}}_0$ under the action of the modified Hamiltonian $\hat{\mathcal{H}} = \hat{\mathcal{H}}_0 + \hat{\mathcal{H}}_1$, where the counterdiabatic term is given by[65]

$$\hat{\mathcal{H}}_1 = \lambda'(t) \sum_{\mathbf{k}} \frac{1}{2|\vec{a}_{\mathbf{k}}(\lambda)|^2} \psi_{\mathbf{k}}^{\dagger} \left[(\vec{a}_{\mathbf{k}}(\lambda) \times \partial_{\lambda} \vec{a}_{\mathbf{k}}(\lambda)) \cdot \vec{\sigma}_{\mathbf{k}}\right] \psi_{\mathbf{k}}. \tag{24}$$

The auxiliary Hamiltonian $\hat{\mathcal{H}}_1$ involves highly nonlocal pairwise interactions in the fermionic representation and many-body interactions in the spin representation, accessible in quantum simulators.[68–70] If the range of the auxiliary Hamiltonian $\hat{\mathcal{H}}_1$ is restricted to a value M (which is equivalent to include up to M-body spin interactions), an efficient suppression of excitations occurs in modes with $|\mathbf{k}| > 1/M$, as explicitly verified in the 1D quantum Ising model.[65] Simpler forms of the auxiliary Hamiltonian $\hat{\mathcal{H}}_1$ are obtained whenever $\hat{\mathcal{H}}_0$ contains exclusively homogeneous spin interactions,[71] as in the Lipkin–Meshkov–Glick model.[72]

6. The KZM and Transitions between Steady States

As we have noted already in the introduction, experimental tests of the
power law scaling predicted by the KZM are difficult, since the exponent
that governs the dependence of $\hat{\xi}$ on τ_Q is usually fractional, and often
much less than 1. It is therefore no surprise that the earliest experiments
that were devised to test the KZM scaling were carried out in transitions
between distinct nonequilibrium steady states (rather than between different equilibria) in driven systems,[73–76] where implementing the quench is
often easier. In such systems — for example, in Rayleigh-Bénard convection — the broken symmetry can be associated with convective flows driven
by thermal gradients in the presence of an external potential (e.g. gravity).
Topological defects are the imperfections in the arrangement of these far
from equilibrium convective patterns. For example, the lattice of normally
hexagonal Bénard cells may exhibit lattice defects.

The effective field theory (such as a suitable version of the Ginzburg–
Landau model) is often used to represent symmetry breaking associated
with the formation of such steady-state structures. One can therefore expect
(based on this Ginzburg–Landau connection) that some of the features of
the dynamics of symmetry breaking predicted by the KZM for equilibrium
phase transitions can be also detected in the transitions between distinct
nonequilibrium steady states that exhibit different symmetries. This was
indeed the case in the nonlinear optical system.[73] However, more recent
experiments (see e.g. Ref. 77) present a richer and more complicated picture.
Indeed, the very nature of such steady state phenomena (e.g. the fact that
defects appear in an order parameter defined by the lattice of relatively
large, Bénard cell sized structures) suggests caution in applying the KZM
to transitions between distinct nonequilibrium states that exhibit different
broken symmetries. The concepts such as "the relevant speed of sound" or
the "sonic horizon" and, especially, the ideas underlying renormalization
group (that are natural in the equilibrium second-order phase transitions,
where the KZM was developed) are not directly applicable to switching
between distinct nonequilibrium steady states.

This inapplicability of renormalization is not a concern in the thermodynamic or quantum phase transitions where many orders of magnitude
usually separate, e.g. the healing length, from the microscopic scales that
determine the basic physics. This scale separation allows for the indepen-
dence of the physics that governs dynamics of the order parameter (and,
hence, e.g. the size of the sonic horizon) from the underlying microphysics.

However, when one cannot appeal to renormalization, scalings deduced from the KZM need not hold, or could be only an approximation.

An interesting and instructive recent example of the extent to which the KZM can be used as a guide in such more general class of symmetry breaking phenomena even when the underlying dynamics does not yield itself to renormalization (or, indeed, to modeling of the order parameter in terms of partial differential equations) is offered by experiments[78] and computer simulations.[78,79] In this case what happens is in qualitative agreement with KZM, but does not follow its predictions in detail. A quantum example of an oversimplified model of the Bose–Einstein condensation that did seem to approximately follow the KZM even though the usual BEC order parameter did not enter the discussion, and the dynamics was represented by transitions between discrete states — as in Ref. 79 — was also analyzed some time ago.[80]

In spite of these caveats, the transitions between steady states have provided suggestive early evidence of KZM "mean field" scalings.[73] Recent interesting work (see Refs. 78 and 79, and references therein) can be regarded as an attempt to formulate an extension of KZM that might be, possibly in only an approximate way, valid even where there is no scaling traceable to renormalization, and even where partial differential equations cannot be used to represent bifurcation-like processes under study.

7. Winding Numbers in Loops

The earliest prediction[19] of scaling of the topologically nontrivial configurations induced by phase transition dynamics concerned winding numbers (and the resulting flows) in annular superfluid containers, see Fig. 3.

The basic reasoning is straightforward: consider an annulus of circumference $\mathcal{C}$ that contains a substance which, as a result of a change in the external parameters, becomes a superfluid (or superconductor). When the characteristic healing length set by the phase transition dynamics is $\hat{\xi}$, and $\mathcal{C} \gg \hat{\xi}$ while the width of the annulus is small so that it can be regarded as effectively one-dimensional loop, there will be

$$N \simeq \frac{\mathcal{C}}{\hat{\xi}} \tag{25}$$

sections of the annulus that independently select the phase of the condensate wavefunction. As a consequence of the resulting random walk of phase the typical net phase mismatch accumulated over the length $\mathcal{C}$ of the loop

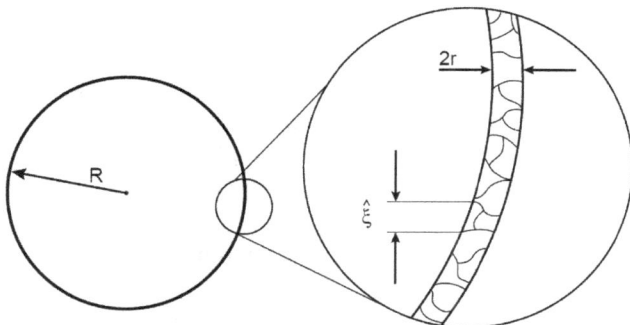

Fig. 3. When a superfluid transition occurs in an annular container with the width comparable to the frozen-out healing length $\hat{\xi}$, distinct domains will choose the phase of the superfluid wavefunction independently. There will be then $N \simeq \mathcal{C}/\hat{\xi}$ of such domains with randomly chosen phase, or, in other words, along the circumference $\mathcal{C} = 2\pi R$ a phase mismatch $\Delta\Theta \simeq \sqrt{\mathcal{C}/\hat{\xi}}$ will appear as a consequence of such random walk. The resulting phase gradient implies that a quantized, persistent flow can be induced by the KZM in a superfluid transition.

will be given by[19]

$$\Delta\Theta \simeq \sqrt{N} \simeq \sqrt{\frac{\mathcal{C}}{\hat{\xi}}} \,. \tag{26}$$

This net phase mismatch implies an average winding number:

$$\mathcal{W} \simeq \frac{\Delta\Theta}{2\pi} \,. \tag{27}$$

After the phase ordering has smoothed out the domains, the resulting superfluid will flow with the velocity given by the phase gradient:

$$v = \left| \frac{\hbar}{m} \vec{\nabla}\Theta \right| \simeq \frac{\hbar}{m} \sqrt{\frac{1}{\mathcal{C}\hat{\xi}}} \,. \tag{28}$$

In the case of superconductors similar reasoning[23] leads to magnetic field trapped inside $\mathcal{C}$ corresponding to the number of quanta given by $\mathcal{W}$.

The basic idea of a random walk in phase resulting in the nonzero winding number has been successfully tested in the experiment involving a loop that was deliberately divided into $N = 214$ superconducting sections by "weak links".[7] When the loop was reconnected into a single superconducting ring, flux quanta were trapped inside. Over many runs of this experiment, the resulting quantized magnetic flux had an approximately Gaussian distribution with the dispersion (related to the typical winding number) well approximated by the KZM-like Eqs. (26) and (27).

By the very nature of the above reconnection experiment, the quench rate of the transition was irrelevant, and, indeed, not well defined: the size of the "domains" that can choose the same phase was set by the distance between the weak links, so that the number of such domains was constant ($N = 214$), and hence independent of the quench rate.

The dependence of the typical trapped winding numbers on the quench rate is difficult to test in the laboratory. The expected power law is even smaller (by a factor of 2) than the already small fractional power $\frac{\nu}{1+z\nu}$ that governs the size of $\hat{\xi}$. Moreover, the quench should be uniform — it must happen nearly simultaneously in the whole annulus — for, otherwise, the speed of the relevant sound may exceed the speed of the transition front, so the regions that "go superfluid" first will communicate their choice of the phase selection to the neighborhood, and the resulting winding numbers can be suppressed.[21,22]

Numerical simulations of the stochastic Gross–Pitaevskii equation,[29] such as those in Fig. 4, confirm this general paradigm and verify the KZM-predicted scalings. They also show how sensitive the resulting winding number is to the imperfections in the implementation of the transition. Such difficulties have so far hampered experimental verification of the KZM-predicted winding number scaling with the transition rate in, e.g. gaseous BECs.

7.1. *Trapping flux in small loops*

An interesting and successful set of increasingly sophisticated experiments that yielded a power law was carried out in small superconducting systems with the topology of an annulus: tunnel Josephson junctions and small superconducting loops.[81–84] In this regime $\mathcal{C} \ll \hat{\xi}$, so that the winding numbers other than $\mathcal{W} = 0, \pm 1$ are exceedingly unlikely, and the natural observable in this case is the frequency of trapping a winding number $|\mathcal{W}| = 1$.

As the random walk takes no more than one step, the square root of Eq. (26) is no longer relevant, and it is reasonable to expect changes in the power law scaling with the quench rate. This general conclusion was reached using field-theoretic methods to predict doubling of the power law compared with the $\mathcal{C} \gg \hat{\xi}$ regime.[85] Thus, when $\hat{\xi} \sim \tau_Q^{\frac{1}{4}}$ (as is expected

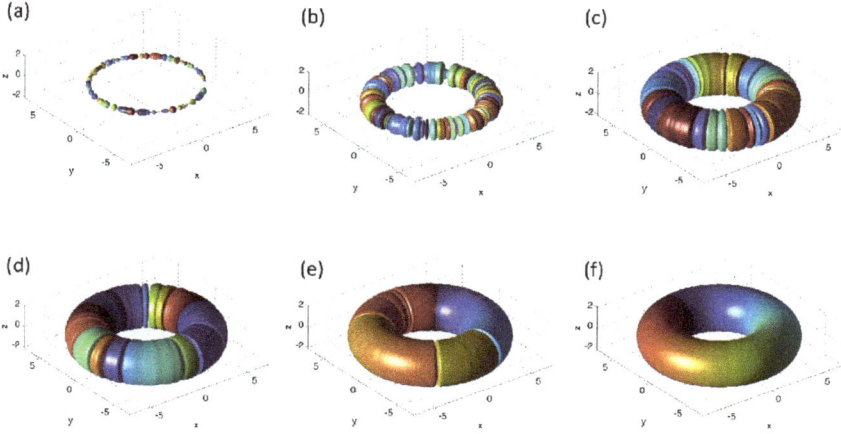

Fig. 4. Sequence of snapshots of isodensity surfaces during the growth of a BEC in a toroidal trap resulting in the formation of a superfluid current, modeled by the stochastic Gross–Pitaevskii equation.[29] The color coding describes the phase of the condensate along the ring. An early stage is characterized by large density and phase fluctuations. As the condensate grows there is a coarsening of both phase and density fluctuations that result in the appearance of solitons. The final estate exhibits a uniform density and winding number $\mathcal{W} = 1$.

in low-temperature superconductors), Ref. 85 predicted exponent of $\frac{1}{4}$ for the scaling of typical winding numbers when $\mathcal{C} \ll \hat{\xi}$ (as opposed to the exponent $\frac{1}{8}$ valid for $\mathcal{C} \gg \hat{\xi}$, Eq. (26)).

Initial experiments[81] yielded the power law of the frequency of trapping a fluxon consistent with this prediction. However, later (and more refined and presumably more accurate) experiments[83] resulted in a steeper slope with the exponent close to 0.5 as shown in Fig. 5, i.e. twice the prediction of Ref. 85. This discrepancy was puzzling. Moreover, experiments on small superconducting loops by Monaco et al.[86] reported similar scaling of the frequency of trapping with the exponent of 0.62 ± 0.15. This exponent is consistent with 0.5, again four times the slope expected for the scaling of typical $\mathcal{W}$ in the $\mathcal{C} \gg \hat{\xi}$ regime. The discrepancy with the initially anticipated scaling[85] in the tunnel Josephson junctions was attributed to the possible fabrication problems and the resulting "proximity effect".[83]

The resolution of the mystery that does not call on fabrication problems and resulting complications may be assisted by the recent observation[87] that in the regime of small loops, $\mathcal{C} \ll \hat{\xi}$, dispersion of the winding numbers

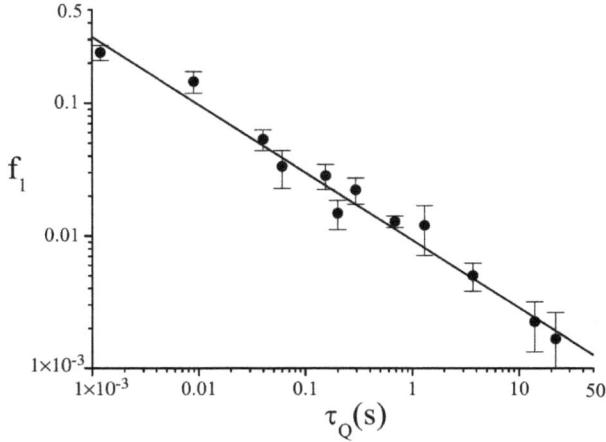

Fig. 5. Scaling of the frequency f_1 of trapping a single fluxon with winding number $|\mathcal{W}| = 1$ in annular Josephson tunnel junctions as a function of the quench rate τ_Q. Each point is the result of averaging over many thermal cycles. A fit to a power law $f_1 = a\tau_Q^{-\alpha}$ leads to $\alpha = 0.51$. From Monaco et al.[83] © 2006 American Physical Society.

$\sqrt{\langle \mathcal{W}^2 \rangle}$ scales differently than $\langle |\mathcal{W}| \rangle$. Indeed, $\langle |\mathcal{W}| \rangle$ scales as probabilities (and, hence, frequencies) of $|\mathcal{W}| = 1$ while the dispersion scales as a square root of that probability. Therefore, it appears to us that the prediction of doubling of the scaling exponent of Ref. 85 is relevant to the dispersion $\sqrt{\langle \mathcal{W}^2 \rangle}$, while in the experiments that measure frequency of detection of $|\mathcal{W}| = 1$ one should expect four times the slope of the dispersion in the large-loop regime, $\mathcal{C} \gg \hat{\xi}$. With this revision[87] of the original expectations,[85] the experiments on tunnel Josephson junctions as well as on the small superconducting loops are in excellent agreement with the predictions of the KZM and can be regarded as its verification (albeit in the mean field case).

The reason for the quadrupling (rather than just a simple doubling) of the power law for frequencies as well as for typical winding numbers characterized by $\langle |\mathcal{W}| \rangle$ is straightforward.[87] We first note that the charges of topological defects created by the quench are anticorrelated.[88] This is reflected in the Eqs. (26) and (27) that recognize the phase of the condensate as the fundamental random variable. By contrast, if charges were assigned at random, typical $\mathcal{W}$ would be given by the square root of the number of defects subtended by the circumference $\mathcal{C}$. Thus, for $\mathcal{C} \gg \hat{\xi}$, when the contour contains many defects of both charges, random distribution of charges would be directly proportional to the circumference (rather than its square root, Eq. (26)).

This scaling of typical $\mathcal{W}$ with $\mathcal{C}$ can be recovered in a simple model[87] where pairs of the oppositely charged defects are randomly scattered on a plane (see Fig. 6). The sizes of the pairs, as well as their separations, are

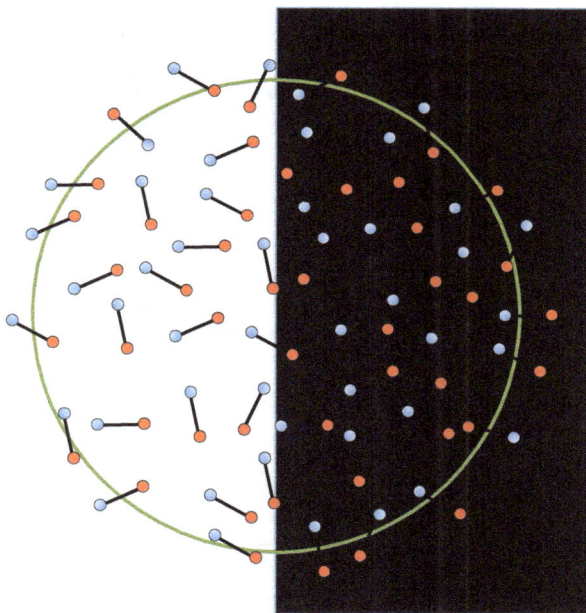

Fig. 6. The winding number for the circumference $\mathcal{C}$ (the circle above) due to vortex-antivortex pairs scattered randomly on a plane. Contribution of the pairs that are completely outside or completely inside $\mathcal{C}$ vanishes: only pairs that straddle the contour contribute to $\mathcal{W}$. The number of such pairs is proportional to the circumference $\mathcal{C}$. Note that pairing illustrated above is in a sense imaginary (as is suggested by the right-hand side of the figure, where pair assignments are invisible), as there is generally no unique "correct" way to combine vortices and antivortices into pairs. Nevertheless, the recognition of pairing leads to correct scaling of winding numbers with $\mathcal{C}$. When loops are so small that typically, at most only one end of a pair "fits inside", scaling changes, see Fig. 7. From Zurek.[87] © IOP Publishing. Reproduced by permission of IOP Publishing. All rights reserved.

presumably of the order of $\hat{\xi}$. The typical winding number is then given by the square root of the number of pairs dissected by $\mathcal{C}$, which leads to the scaling of Eq. (26).

The heuristic picture of the generation of the winding number is then straightforward. The quench results in a random configuration of the order parameter deposited inside $\mathcal{C}$. Instead, we can imagine an infinite plane with all configurations of the order parameter left behind by the transition pockmarked by defects and sampled at random by dropping the contour $\mathcal{C}$ at random locations. When defects are paired up (as their anticorrelations suggest) and $\mathcal{C} \gg \hat{\xi}$, the scaling of Eq. (26) is easily recovered (see Fig. 7). By contrast, when $\mathcal{C} \ll \hat{\xi}$, most of the loops tossed on the plane will end

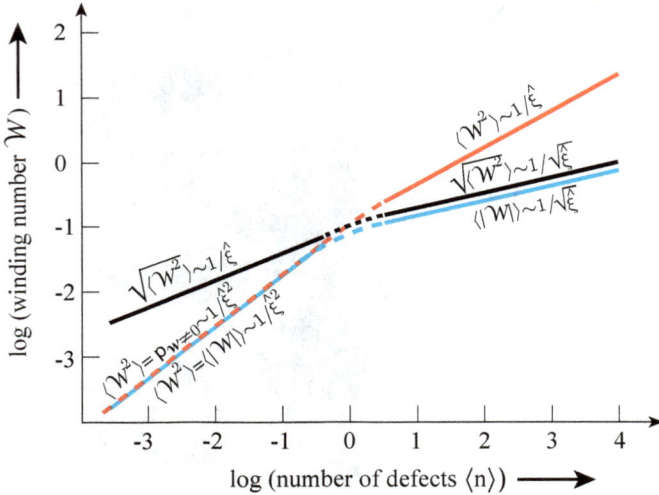

Fig. 7. The tilt of the scaling of the dispersion, $\sqrt{\langle \mathcal{W}^2 \rangle}$, its square, and the average $\langle |\mathcal{W}| \rangle$ of the winding numbers is expected to change with the number of defects trapped inside $\mathcal{C}$ when $\langle n \rangle \simeq 1$.[87] For loops that trap many defects, $\langle n \rangle \gg 1$, the dispersion and the average absolute value of $\mathcal{W}$ scale similarly, $\sqrt{\langle \mathcal{W}^2 \rangle} \sim |\mathcal{W}| \sim \sqrt{\mathcal{C}/\hat{\xi}}$. However, different tilts, corresponding to the exponents that control the slopes of the dispersion and $\langle |\mathcal{W}| \rangle$, set in as $\langle n \rangle \ll 1$. Compared to $\sqrt{\mathcal{C}/\hat{\xi}}$, the slope of the dispersion *doubles*, $\sqrt{\langle \mathcal{W}^2 \rangle} \sim \sqrt{A_{\mathcal{C}}}/\hat{\xi}$ while the slope of the average absolute value *quadruples* so that $\langle |\mathcal{W}| \rangle \simeq p_{|W|=1} \sim A_{\mathcal{C}}/\hat{\xi}^2 \sim \langle \mathcal{W}^2 \rangle$ when $\langle n \rangle \ll 1$, where $A_{\mathcal{C}}$ is the area enclosed by the contour $\mathcal{C}$. (Note that $\langle n \rangle \approx A_{\mathcal{C}}/\hat{\xi}^2$). From Zurek.[87] © IOP Publishing. Reproduced by permission of IOP Publishing. All rights reserved.

up "empty", hence, will have $\mathcal{W} = 0$. Only on rare occasions when the loop of area $A_{\mathcal{C}} \sim \mathcal{C}^2 \ll \hat{\xi}^2$ "traps" a defect inside, the winding number will be $+1$ or -1, depending on the defect charge. Moreover, the probability of trapping the defects will scale as $\sim A_C \times \hat{\xi}^{-2}$, as the density of the KZM defects is $\sim 1/\hat{\xi}^2$. Consequently, the probability (and, hence, the frequency) of finding a loop with $|\mathcal{W}| = 1$ in the case of $\langle|\mathcal{W}|\rangle \ll 1$ scales as:

$$p_{\mathcal{W}=\pm 1} \sim \frac{A_C}{\hat{\xi}^2} \sim \frac{C^2}{\hat{\xi}^2} \ . \tag{29}$$

Note that the power with which $\hat{\xi}$ appears above is *four* times the power law relating typical $\mathcal{W}$ and $\hat{\xi}$ when $C \gg \hat{\xi}$, Eq. (26). It follows that the scaling of the frequency (or of $\langle|\mathcal{W}|\rangle$) with τ_Q in this $C \ll \hat{\xi}$ regime is four times steeper than for large $(C \gg \hat{\xi})$ loops. Scaling of $\langle\mathcal{W}^2\rangle$ in this regime is equally steep, as $\langle\mathcal{W}^2\rangle \approx \langle|\mathcal{W}|\rangle \approx p_{\mathcal{W}=\pm 1} \simeq C^2/\hat{\xi}^2$. On the other hand, scaling of $\sqrt{\langle\mathcal{W}^2\rangle}$ will only double (which is what may have been predicted by Ref. 85). It is that discrepancy between the scaling of dispersion and frequency of detection in case of small loops that may account for the experimental results seen in Fig. 5.

This quadrupling is a combination of two doublings (or, rather, it reverses the consequences of two square roots that appear as the size of the loop increases from $C \ll \hat{\xi}$ to $C \gg \hat{\xi}$). For small loops, the frequency of trapping a single defect is proportional to the area inside C, and this yields a proportionality to the area for $\langle\mathcal{W}^2\rangle \approx \langle|\mathcal{W}|\rangle \approx p_{\mathcal{W}=\pm 1} \simeq C^2/\hat{\xi}^2$. By contrast, for large loops the net winding number is given by the random walk in the phase (which yields square root #1) of the number of pairs intercepted by $C \gg \hat{\xi}$ rather than the area inside, A_C (which implies square root #2). This change of the power law that governs the scalings will be reflected in the power law dependence of the winding number on the quench time τ_Q.

8. Defect Formation in Multiferroics

Multiferroics are materials that exhibit more than one primary ferroic order parameter simultaneously (i.e. in a single phase). Recent measurements in rare earth multiferroics have provided what may be a compelling evidence of the KZM.[89] The reason for excitement is illustrated in Fig. 8. It shows snapshots of the surface of $ErMnO_3$ cooled, at different rates, from about 1200°C (i.e. from above the phase transition that occurs at 1120–1140°C) to room temperature. The mosaic pattern seen in this figure represents domains that form as a result of symmetry breaking. These domains are punctuated by vortex-like defects that appear where several domains meet. The topological charge of the point defect is determined by the order in which distinct phases are arranged. Clearly, the scale of the structures (that can be deduced from the density of the point defects) increases with the cooling time, as is shown in Fig. 8.

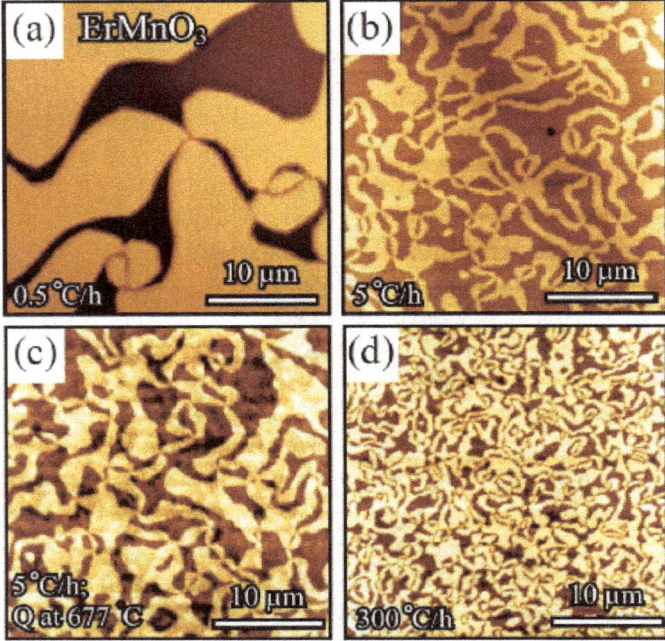

Fig. 8.   Vortices (present where three dark and three bright domains merge) punctuate domain patterns formed in chemically-etched ErMnO$_3$ crystals for different cooling rates. The characteristic scale extracted from the experiments (e.g. from the density of vortices) exhibits a scaling with the rate of quench that is consistent with the KZM predictions for the 3D XY model,[90] which has $\nu = 0.6717$ and $\nu z = 1.3132$ calculated using Monte Carlo simulations.[91] From Chae *et al.*[89] © 2012 American Physical Society.

The power-law exponent governing the dependence of the distance between the point defects in the quench rate is close to 0.25, which suggests a description in terms of the KZM with, e.g., the mean-field critical exponents $\nu = \frac{1}{2}$, $z = 2$ that would result in $\hat{\xi}$ scaling with the power $\frac{\nu}{1+\nu z} = \frac{1}{4}$. However, Griffin *et al.*[90] note that the universality class of the phase transition is the same as of the 3D XY model, which has $\nu = 0.6717$ and $\nu z = 1.3132$ calculated using Monte Carlo simulations[91] (as a caveat, note that the choice of z depends on the dynamics, which is not well characterized in this case). Consequently, the predicted KZM exponent would be $\frac{\nu}{1+\nu z} \simeq 0.29$. This is consistent with the experimentally measured value. Indeed, not just the slope of the power law dependence but also the net defect density are in approximate agreement with the *ab initio* calculations.[90]

While the above discussion can be interpreted as a resounding confirmation of the KZM, there are reasons for caution. To begin with, apart from the approximate critical temperature, very little is known about the actual critical behavior of $ErMnO_3$ (and similar rare earth manganites given by the chemical formula $RMnO_3$, where R stands for a rare earth element). This problem is largely due to the very high critical temperature, which makes, at least to date, the measurements that would allow one to extract ν and z all but impossible. Indeed, at present it is not even completely clear, experimentally, whether the transition is of first or second order.

The other reason for caution comes from the fast temperature quenches carried out recently.[90] They have yielded (albeit in $YMnO_3$, and not in $ErMnO_3$ where the original study[89] was conducted) a surprise: the increase in the rate of much faster quenches (with cooling rates of up to $10^2 K/s$) actually suppressed defect production, resulting in increasing size of domain structures. This has not yet been explained, although several possibly relevant effects have been discussed.[90] At present, it seems reasonable to wait for experimental confirmation of this 'anti-KZM' effect before attempting to advance a detailed theory.

One might hope that the experimental difficulties and the related uncertainties may be eventually overcome. Precision measurements of the scaling of $\hat{\xi}$ with τ_Q could be then increased sufficiently so that one might confirm that it is indeed close to 0.29 predicted by the 3DXY universality class (and not "mean field"), and that would be a major coup.

9. The Inhomogeneous Kibble–Zurek Mechanism

Tests of the Kibble–Zurek mechanism in the laboratory often face the situation in which the phase transition is inhomogeneous as opposed to being crossed everywhere at once.[92] What survives from the Kibble–Zurek mechanism in inhomogeneous phase transitions is decided by causality.[93] This realization has provided a foundation for an active area of research in recent years, with theory works[22,35,93–101] accompanied by a substantial experimental progress[13–16] following the pioneering suggestion by Kibble and Volovik,[21] who first focused on the problem of phase ordering behind a propagating front of a continuous phase transition. This situation can arise as a result of an inhomogeneous tuning of the control parameter driving the transition $\lambda = \lambda(x, t)$. Alternatively, it might result from a spatial dependence of the critical value $\lambda_c = \lambda_c(x)$ which often occurs in trapped systems. Using the local density approximation, one could then replace the critical value $\lambda_c[\rho]$ for homogeneous density ρ by $\lambda_c[\rho(x)]$. We refer the reader to the recent review for a detailed exposition of the subject.[92]

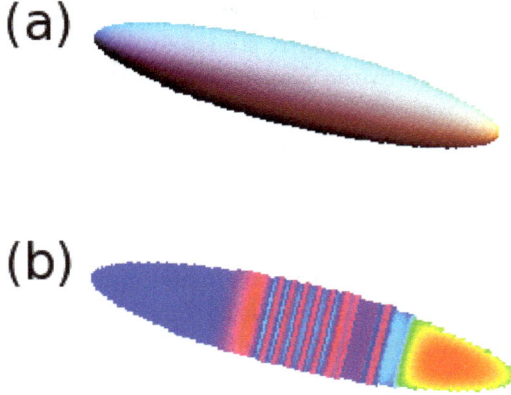

Fig. 9.   Bose–Einstein condensation by evaporative cooling in a harmonic trap offers an example of the inhomogeneous phase transition where causality enhances the dependence of the defect density on the quench rate. (a) Isodensity surface of Bose gas in a trap. Its density is highest in the center of the trap, and that is where the condensation will start when the cloud cools e.g. by evaporative cooling. When the region that becomes BEC first can communicate its choice of the condensate phase to the neighbouring domains, defects will not form. (b) On the other hand, when the speed of relevant sound $\hat{s} = \hat{\xi}/\hat{t}$ is less that the speed v_F with which the critical point propagates as a result of cooling, different phases will be chosen by different domains (as indicated by the color coding) and grey solitons can be created. From Zurek.[93] © 2009 American Physical Society.

Let us assume that the critical point exhibits a spatial dependence $\lambda_c = \lambda_c(x)$ and that the system undergoes a homogeneous quench of the control parameter with constant rate $1/\tau_Q$,

$$\lambda(t) = \lambda_c \left(1 - \frac{t}{\tau_Q}\right), \tag{30}$$

during the time interval $t \in [-\tau_Q, \tau_Q]$. As in the homogeneous case, it is convenient to introduce the reduced control parameter

$$\varepsilon(x, t) = \frac{\lambda_c(x) - \lambda(t)}{\lambda_c(x)}, \tag{31}$$

which takes values $\varepsilon(x, t) < 0$ in the high symmetry phase where the system is initially prepared, it reaches $\varepsilon(x, t) = 0$ at the critical point, and the broken-symmetry phase for $\varepsilon(x, t) > 0$.

To establish the relationship with the homogeneous case, it is further convenient to introduce an effective local quench time,

$$\tau_Q(x) = \left| \frac{\partial \varepsilon(x, t)}{\partial t} \right|^{-1}. \tag{32}$$

The condition $\varepsilon(x_F, t_F) = 0$ allows us to find the time t_F at which the propagating front crosses the transition at the location $x_F = x$

$$t_F(x) = \tau_Q \left[1 - \frac{\lambda_c(x)}{\lambda_c(0)} \right], \tag{33}$$

in terms of which the reduced control parameter reads $\varepsilon(x, t) = \frac{t - t_F(x)}{\tau_Q(x)}$. Matching, in the spirit of Eq. (2),

$$\tau[\varepsilon(x, t)] = \left| \frac{\varepsilon(x, t)}{\dot{\varepsilon}(x, t)} \right| = |\varepsilon(x, t)| \tau_Q(x), \tag{34}$$

one obtains[93,99] that

$$\hat{\varepsilon}(x) = \left[\frac{\tau_0}{\tau_Q(x)} \right]^{\frac{1}{1 + \nu z}}. \tag{35}$$

See Ref. 92 for alternative derivations. We note that $\hat{\varepsilon}(x)$ is associated with the local freeze-out time $\hat{t}(x) = [\tau_0 \tau_Q(x)^{\nu z}]^{\frac{1}{1 + \nu z}}$ measured with respect to $t_F(x)$ (that is, freeze-out will take place in the interval $[t_F(x) - \hat{t}(x), t_F(x) + \hat{t}(x)]$). It follows that the typical size of the domains in the broken symmetry phase is given by

$$\hat{\xi}(x) \equiv \xi[\hat{\varepsilon}(x)] \simeq \xi_0 \left[\frac{\tau_Q(x)}{\tau_0} \right]^{\frac{\nu}{1 + \nu z}}. \tag{36}$$

In contrast to the homogeneous scenario, defect formation does not occur all over the spatial extent L of the system but it is restricted by causality.[93] Once a choice of a ground-state of the vacuum manifold is made locally in a given part of the system, it can be communicated to neighbouring regions. The characteristic local velocity of the perturbations, which determines the speed at which this choice can be communicated, is given by the analogue of the second-sound velocity in ^{4}He that can be upper bounded by the ratio of the local frozen-out correlation length $\hat{\xi}(x)$ and relaxation time scale $\hat{\tau}(x) = \tau[\hat{\varepsilon}] = \hat{t}(x)$, this is, by[21]

$$\hat{s} = \frac{\hat{\xi}}{\hat{\tau}} = \frac{\xi_0}{\tau_0}\left[\frac{\tau_0}{\tau_Q(x)}\right]^{\frac{\nu(z-1)}{1+\nu z}}. \tag{37}$$

When $\hat{s}$ is larger that the transition front velocity v_F, defect formation is suppressed. The speed of propagation of this front can be estimated to be[93]

$$v_F = \left|\frac{dx_F}{dt_F}\right| = \frac{\lambda_c(0)}{\tau_Q}\left|\frac{d\lambda_c(x)}{dx_F}\right|^{-1} = \left|\frac{d\tau_Q(x)}{dx_F}\right|^{-1}. \tag{38}$$

This expression diverges for homogeneous system, or where the system is locally homogeneous (e.g. whenever $\lambda_c(x)$ reaches an extremum). For defects to be formed, $\hat{s} < v_F$ is required.

Numerical and analytical tests have confirmed this intuition, and thus, the role of causality in defect formation both in classical[22,35,98,100] and quantum systems.[96,97] This inequality is generally satisfied in a fraction of the system $\hat{X} = \hat{x}/L$, with $\hat{x} = \{x|v_F > \hat{s}\}$. Within $\hat{x}$, the number of defects can be estimated using $\hat{\xi}(x)$. The resulting density of defects in the whole system is then simply given by the total number of defects formed with the homogeneous density in regions where $v_F > \hat{s}$ divided by the total system size, which in the 1D case reduces to

$$n \simeq \frac{1}{L}\int_{\{x|v_F>\hat{s}\}}\frac{1}{\xi_0}\left[\frac{\tau_0}{\tau_Q(x)}\right]^{\frac{\nu}{1+\nu z}}dx. \tag{39}$$

This expression does not generally lead to a power-law in the quench rate.[99] A power-law does however result in limiting cases.[93,99] Whenever $\lambda_c(x)$ attains an extremum, say at $x = 0$, it can be linearized as

$$\lambda_c(x) = \lambda_c(0) + \frac{\lambda_c''(0)}{2}x^2 + \mathcal{O}(x^3), \tag{40}$$

and the front velocity simplifies to

$$v_F \simeq \frac{\lambda_c(0)}{\tau_Q|x\lambda_c''(0)|}, \tag{41}$$

which diverges at the origin $x = 0$. The effective region of the system where defect formation is allowed by causality can be then estimated by comparing (37) and (41). Assuming $\hat{x}$ is simply connected and small enough, so that $\tau_Q(x) \approx \tau_Q(0)$ within $\hat{x}$, it is found that

$$|\hat{x}| \simeq \frac{\lambda_c(0)}{|\lambda_c''(0)|\xi_0}\left[\frac{\tau_0}{\tau_Q(0)}\right]^{\frac{1+\nu}{1+\nu z}}, \tag{42}$$

which increases with the quench rate, as expected. This results in the total density of defects

$$n \simeq \frac{1}{L} \frac{\lambda_c(0)}{|\lambda_c''(0)|\xi_0^2} \left[\frac{\tau_0}{\tau_Q(0)} \right]^{\frac{1+2\nu}{1+\nu z}}, \tag{43}$$

with a new power-law exponent[93] which is a multiple, by a factor $\frac{1+2\nu}{\nu}$, of what is predicted for the density (e.g. $\hat{\xi}^{-1}$) by the homogenous KZM in 1D, Eq. (10). This constitutes a testable prediction of the inhomogeneous KZM (IKZM). Numerical evidence supporting this scaling was first described in Refs. 35 and 98, see as well Ref. 100. A flurry of experimental activity testing the IKZM has been reported during 2012 and 2013 on the scaling of defect formation in inhomogeneous system and we now turn our attention to it.

10. Kink Formation in Ion Chains

Coulomb crystals made of ion chains stand out as a platform for quantum technologies as a result of their potential for quantum information processing[102] and quantum simulation.[103,104] Coulomb crystals with several millions of ions have been observed both in Paul and Penning traps.[105,106] When the inter-ion spacing a is homogeneous, different structural phases can be accessed by tuning the transverse harmonic confinement. As the trapping frequency ν_t is reduced from high to lower values, the Coulomb crystal undergoes a series of structural phase transitions with phases characterized by linear, zigzag, helicoidal, and more complex structures.[107] These transitions are generally of first order, with the following exception: the transition between the linear ion chain and the doubly-degenerate zigzag phase, shown in Fig. 10(a), is of second order[108-110] and occurs at the critical frequency

$$\nu_{t,c}^2 = \frac{7}{2}\zeta(3)\frac{Q^2}{ma^3}, \tag{44}$$

where $\zeta(p)$ is the Riemann-zeta function and m and Q are the mass and charge of the ions, respectively. A finite-time crossing of this transition is expected to result in the formation of topological defects as described by the KZM,[35,98] see Fig. 10(b).

The axial confinement in an ion chain makes the inter-ion spacing spatially dependent, $a = a(x)$, as illustrated in Figs. 10(c) and 10(d). Using the local density approximation, away from the chain edges the linear density of ions given by the inverse of the distance between them, $a(x)^{-1}$, is well

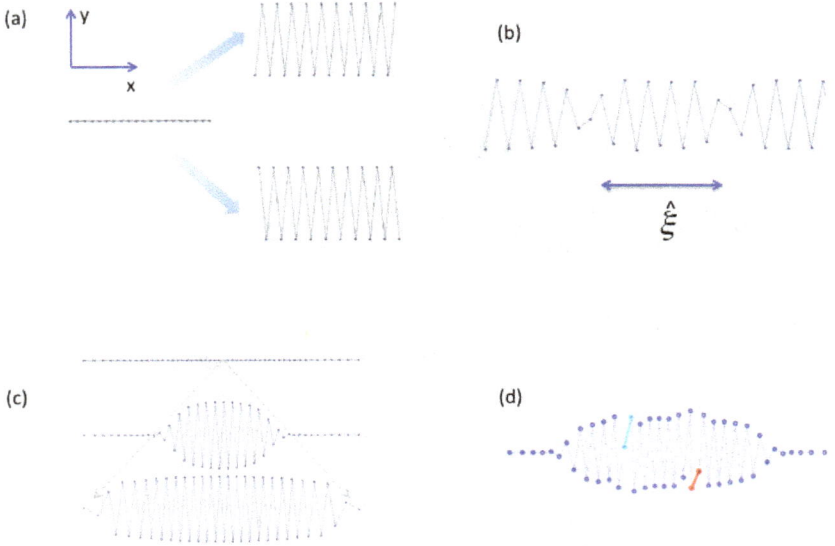

Fig. 10. Linear to zigzag phase transition in ion chains. (a) Symmetry breaking in an homogeneous ion chain following a decompression of the transverse confinement. The vacuum manifold consists of two doubly-degenerate disconnected regions, associated with a $\mathbb{Z}_2$-symmetry breaking scenario. (b) Boundaries between disparate choices of the vacuum lead to the formation of $\mathbb{Z}_2$-kinks or domain walls. (c) A sequence of ground state configurations in a harmonically trapped ion chain. As a result of the axial harmonic confinement the inter-ion spacing $a(x)$ is lowest at the center of the chain and increases sideways. Under a (homogeneous) decompression of the transverse confinement, the zigzag phase is first formed in the center of the chain (where Coulomb repulsion is higher) and co-exists with the region in the linear phase. (d) The transverse decompression (or axial compression) of an inhomogeneous ion chain at finite rate can lead to the formation of structural defects. These defects are not stationary and can propagate along the chain and annihilate by collisions (between a kink and an anti-kink with opposite topological charge) or can be lost at the edges of the chain.

approximated by the inverted parabola[35,98]

$$a(x)^{-1} = \frac{3}{4}\frac{N}{L}\left(1 - \frac{x^2}{L^2}\right), \tag{45}$$

where N is the number of ions, L is half the length of the chain, and x the distance from the center. This leads to a spatial modulation of the critical frequency along the chain,

$$\nu_{t,c}(x)^2 = \frac{7}{2}\zeta(3)\frac{Q^2}{ma(x)^3} , \tag{46}$$

which ultimately makes the linear to zigzag transition inhomogeneous. In the thermodynamic limit, the system obeys an effective time-dependent Ginzburg–Landau equation where the difference $\nu_t^2 - \nu_{t,c}^2$ governs the transition from the high-symmetry phase to the broken symmetry phase.[35,98]

Consider a quench of the transverse trap frequency ν_t, such that

$$\nu_t(t)^2 = \nu_{t,c}(0)^2 - \delta_0^2 \frac{t}{\tau_Q} . \tag{47}$$

Around the critical point the transverse frequency can be linearized, $\nu_t \approx \nu_{t,c}(0) - \delta \frac{t}{\tau_Q}$ with $\delta = \delta_0^2/[2\nu_{t,c}(0)]$. Under such a quench, as a result of the spatial dependence of $\nu_{t,c}(x)$, the zigzag phase is not formed everywhere at once, and it arises first in the center of the chain. To account for the formation of kinks it is required to extend the KZM to inhomogeneous scenarios as in Sec. 9, see Refs. 35 and 98 in this context. The reduced squared-frequency

$$\varepsilon(x,t) = \frac{\nu_{t,c}(x)^2 - \nu_t(t)^2}{\nu_t^c(x)^2} \tag{48}$$

governs the divergence of the correlation length and the relaxation time at the critical point

$$\xi = \frac{\xi_0}{\sqrt{\varepsilon(x,t)}}, \qquad \tau = \frac{\tau_0}{\sqrt{|\varepsilon(x,t)|}} , \tag{49}$$

where ξ_0 and τ_0 are set by the inter-ion spacing $a(0) = a$ and the inverse of the characteristic Coulomb frequency is given by $\omega_0^{-1} = \sqrt{ma^3/Q^2}$. We have assumed that the system is underdamped which is the case whenever the dissipation strength η induced by laser cooling satisfies $\eta^3 \ll \delta_0^2/\tau_Q$. This leads to the critical exponents $\nu = 1/2$, and $z = 1$. The front crossing the transition satisfies $\varepsilon(x,t) = 0$ and reaches x at time

$$t_F(x) = \tau_Q \left(1 - \frac{\nu_{t,c}(x)^2}{\nu_{t,c}(0)^2}\right). \tag{50}$$

Relative to this time, it is possible to rewrite the reduced squared-frequency

$$\varepsilon(x,t) = \frac{t - t_F(x)}{\tau_Q(x)} , \tag{51}$$

in terms of the local quench rate

$$\tau_Q(x) = \tau_Q \frac{\nu_{t,c}(x)^2}{\nu_{t,c}(0)^2} = \tau_Q(1 - X^2)^{-3}, \tag{52}$$

where $\nu_{t,c}(x)^2 = \nu_{t,c}(0)^2[1 - X^2]^3$ and $X = x/L$. The front velocity reads

$$v_F \sim \frac{\delta_0^2}{\tau_Q} \left| \frac{d\nu_{t,c}(x)^2}{dx} \right|_{x_F}^{-1} = \frac{L\delta_0^2}{6\nu_{t,c}(0)^2\tau_Q} \frac{1}{|X|}(1 - X^2)^{-2}. \tag{53}$$

The essence of the Inhomogeneous Kibble–Zurek mechanism (IKZM) is the interplay between the velocity of the front and the sound velocity.[21,93] As in Sec. 9 (see Eq. (37)), the relevant velocity of perturbations can be estimated to be

$$\hat{s} = \frac{\hat{\xi}}{\hat{\tau}} = \frac{\xi_0}{\tau_0} \left(\frac{\tau_0}{\tau_Q(x)} \right)^{\frac{\nu(z-1)}{1+\nu z}} = a\omega_0. \tag{54}$$

The last equality holds whenever the dynamic critical exponent $z = 1$, as in an underdamped ion chain.

In the IKZM, the condition for kink formation to be possible is given by the inequality

$$v_F(x) > \hat{s}, \tag{55}$$

while the propagation of the pre-selected phase is expected otherwise. As shown in Ref. 99, it is instructive to study the spatial dependence of the ratio $v_F(x)/\hat{s}$, which as a function of $X = x/L$ turns out to be parametrized by the dimensionless quantity $\mathcal{A} = \frac{L\delta_0^2}{6\nu_{t,c}(0)^2 a\omega_0\tau_Q}$. Using the Thomas–Fermi approximation for the axial density, Eq. (45), Fig. 11 shows that typically $v_F(x) > \hat{s}$ is satisfied in two disconnected regions. However, the outer region can be safely disregarded given that kinks possibly formed there are likely to leak out to the linear, outer part of the chain. A kink experiences a Peierls-Nabarro oscillatory potential whose amplitude diminishes with the transverse amplitude of the zigzag (the order parameter), this is, towards the edges of the chain.[98,111] Langevin dynamics simulations show that kinks experiencing a gradient of the zigzag amplitude travel towards the edges of the chain where they disappear. As a result, it suffices to consider the central region of the chain $[-\hat{x}, \hat{x}]$ for defect formation. Generally $\hat{x}$ has to be found numerically. However, when the defect formation is restricted to a region $\hat{X} \ll 1$, then one can set[98]

$$\hat{x} = |\hat{X}|L = \frac{\delta_0^2 L^2}{6\nu_{t,c}(0)^2\tau_Q\hat{v}} + \mathcal{O}(\hat{X}^3). \tag{56}$$

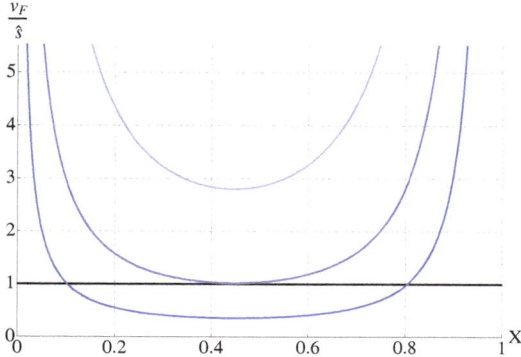

Fig. 11.   Ratio of the front velocity over the second speed of sound as a function of the dimensionless coordinate X measured from the center of the chain ($X = 0$) towards an edge ($X = 1$, in the Thomas–Fermi approximation). The ratio is symmetric around $X = 0$ and only $X > 0$ is displayed for clarity. Within the approximation of the IKZM we adopt, formation of defects with KZM densities is only possible there where $v_F/\hat{s} > 1$. The curves correspond to different values of $\mathcal{A} = \frac{L\delta_0^2}{6\nu_{t,c}(0)^2 a\omega_0\tau_Q}$ with which color coding increases from light to dark blue taking the values $\mathcal{A} = 0.1, 0.289, 0.8$. Above a critical value $\mathcal{A}_c \approx 0.289$ the homogeneous KZM applies. For $\mathcal{A} < \mathcal{A}_c$ domain formation is expected in two disjoint regions and the inhomogeneous KZM governs the dynamics of defect formation. Disregarding the outer region (where defect losses are dominant), whenever the size of the central region is approximately linear in $\mathcal{A}$, the density of defects scales with a power-law in the quench rate.

Under the assumption $\hat{X} \ll 1$, one finds the estimate predicted in Refs. 35 and 98 for the density of kinks

$$n \approx \frac{2\hat{x}}{\hat{\xi}L} = \frac{L}{3\nu_{t,c}(0)^2 a^2\omega_0^2}\left(\frac{\delta_0^2}{\tau_Q}\right)^{4/3}. \tag{57}$$

Note that setting $\tau_Q(x) = \tau_Q$ is consistent with $\hat{X} \ll 1$. We note that without restricting to $\hat{X} \ll 1$, there is no reason to expect a power-law scaling, see Ref. 99.

For sufficiently long quench times, the effective size of a domain set by the KZM length becomes comparable to the (effective) system size, $2\hat{x} = 2\hat{X}L \sim \hat{\xi}$. In this situation, typically one obtains 0 or 1 defects per realization. It was pointed out some time ago in the discussion of the winding numbers trapped in loops (see Sec. 7), that the scaling with τ_Q is likely to steepen[85] when the circumference $\mathcal{C}$ of the loop becomes less than $\hat{\xi}$. This prediction has found support in numerical studies of the dispersion of the winding numbers, $\sqrt{\langle\mathcal{W}^2\rangle}$:[86,112,113] the doubling of the exponent that governs the scaling of the dispersion of $\mathcal{W}$ when $\mathcal{C} \gg \hat{\xi}$ was seen in the $\mathcal{C} \ll \hat{\xi}$ regime, i.e. when $\mathcal{W} = \pm 1$ is much less likely than the probability of $\mathcal{W} = 0$, as $\sqrt{\langle \mathcal{W}^2 \rangle}$ is then dominated by the probability of taking a single step in the random walk, while the contribution of $\mathcal{W} > 1$ is negligible.[87]

We have already seen is Sec. 7 that relating this prediction to experiments requires some care, as doubling of the dispersion of $\mathcal{W}$ with τ_Q in this $\mathcal{C} \ll \hat{\xi}$ regime actually implies *quadrupling* of the frequency of detections of $|\mathcal{W}| = 1$,[87] and the frequency is then the obvious observable. In the case of winding number the situation is relatively well understood, at least at the level of simple models. The quadrupling predicted there (and possibly already observed, see Fig. 5) is in a sense a product of two doublings. One of them comes about as a consequence of the square root that is related to the random walk of the phase that becomes unnecessary in the case when that random walk has only one step. It is likely relevant only in the case of loops. The second doubling has to do with the change of the character of the question: in small loops the focus is trapping a single defect (and the answer is proportional to the area) while in large loops what matters is the number of pairs intercepted by the circumference (and the answer is proportional to the circumference, and, hence, to the square root of the area). It is not clear whether at least one doubling survives the transition from loops to open boundary conditions in the case when the size of the system becomes smaller than $\hat{\xi}$, and excitations becomes rare. Computer experiments with the experimental parameters[15] are consistent with three regimes: KZM (density scaling with power $\sim \frac{1}{3}$), followed by IKZM (density scaling steepening to $\sim \frac{4}{3}$), and, finally — when the density becomes synonymous with the probability of a single kink — an even steeper power law that can be interpreted as $\sim \frac{8}{3}$ of "doubled" IKZM, see Fig. 12. Thus, assuming a doubling of the IKZM this probability can be estimated to be

$$p_1 \sim \left(\frac{2\hat{x}}{\hat{\xi}} \right)^2 \sim \frac{L^4}{\nu_{t,c}(0)^4 a^4 \omega_0^4} \left(\frac{\delta_0^2}{\tau_Q} \right)^{8/3}. \tag{58}$$

Three different experimental groups reported tests of the IKZM in the context of kink formation in trapped ion chains. Experiments[13,15] followed closely the proposal in Refs. 35 and 98, where critical dynamics was driven by a finite-rate decompression of the transverse confinement. The experiment at Mainz[14] used instead a compression along the axial direction. The system sizes and the accessible quench rates in these experiments correspond precisely to the onset of adiabatic dynamics, where $\{0, 1\}$ defects are observed per realization.[14,15] The experiment at Simon Fraser University[13]

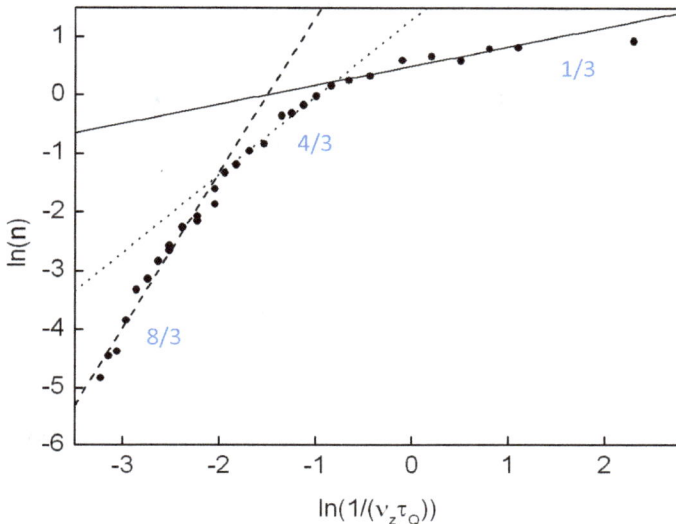

Fig. 12. Scaling of the density of spontaneously formed kinks as a function of the quench rate at which the linear to zigzag structural phase transition is crossed in a trapped Coulomb crystal. The simulations are based on coupled Langevin equations using the experimental parameters in Ref. 15. The system is underdamped in the presence of laser cooling. The plotted lines are guidelines to the eye with slopes predicted by the homogenous KZM (1/3), the IKZM (4/3), and twice the slope predicted by IKZM. Adapted from Ref. 15.

reported broader kink number distributions but the presence of substantial kink losses prevented testing any signature of universality in the dynamics of kink formation. See Table 1 for a summary of these experiments. The results of Refs. 14 and 15 suggest an agreement with Eq. (58).

Table 1. Experimental results on the topological defect formation in ion Coulomb crystals.[13-15] Data was fitted to a power-law in the quench rate τ_Q of the form $n \propto \tau_Q^{-\alpha}$.

Group	Number of ions	Kink number	Fitted exponent α
Mainz University[14]	16	{0, 1}	2.68 ± 0.06
PTB[15]	29 ± 2	{0, 1}	2.7 ± 0.3
Simon Fraser University[13]	42 ± 1	{0, 2}	3.3 ± 0.2

There are obvious concerns about the extent to which the limited data behind Table 1 can be regarded as a verification of the KZM (e.g. the restricted range of quench rates in each regime, the losses of defects, etc.). However, over and above such experimental issues there are two concerns related to the applicability of IKZM theory to the ion chains of the size $\hat{X} \ll \hat{\xi}$ accessible so far in the experiments.[13–15]

Our discussion in Sec. 9 was based on the idea (put forward in the analysis of soliton formation[93]) that a system in an anisotropic harmonic trap can be cleanly divided into regions where the phase transition front velocity is faster (or slower) than the relevant speed of sound. As a consequence, one can distinguish regions where the homogeneous KZM holds (or does not) and defects are created with the local density set by $\hat{\xi}$ (or not at all).

This sharp division is the key assumption underlying the IKZM. However, in computer simulations and analytic studies the transition between the "homogeneous KZM" and "no defects at all" is not completely sharp, and it seems unlikely (e.g. in view of the behavior of the order parameter in the presence of the gradients[94]) that it could be less than $\hat{\xi}$. Thus, the applicability of the IKZM scalings to the ion chains where $\hat{X} \ll \hat{\xi}$ can be questioned at least in the experiments with rather small systems,[13–15] as the limits on the integral in Eq. (58) are not well defined.

The above concern may apply to the IKZM in all small systems. It appears in addition to the difficulties involved in testing the KZM in many-body systems of modest size, where the scaling in the near-critical regime may not have converged to the values of critical exponents that determine the universality class. This concern can be of course addressed by carrying out suitable equilibrium measurements to verify ν and z, and determining that they extend over the range relevant for the KZM.

The KZM is a way to employ equilibrium scalings in predictions of the consequences of nonequilibrium quenches. Checking if the system in question follows the behavior predicted for its equilibrium universality class seems like a prudent first step when the system is of modest size and especially if it is in a trap or any other setting that can potentially complicate its behavior.

The other way to address such concerns is to work with large homogeneous systems. In case of ion traps this is not out of the question: large "racetrack" traps are in principle possible[105] and could be used to study phase transitions and symmetry breaking in ion Coulomb crystals in a setting where the homogeneous KZM could be tested.[114]

10.1. *Prospects of ground-state cooling of ion chains*

In the theory and experiments just discussed, the ion chain is hot enough so that thermal fluctuations dominate over quantum fluctuations, and a classical description applies.

The prospects of achieving ground-state cooling, while experimentally challenging, might pave the way to study the dynamics of quantum phase transitions in ion traps and similar settings. The equilibrium and dynamic properties of the quantum linear to zigzag structural transition have been investigated.[115–119]

Accessing the quantum regime would also pave the way to the experimental realization of topological Schrödinger cats, nonlocal quantum superpositions of conflicting choices of the broken symmetry or quantum phases of matter.[120] Superposition of macroscopic states have been also explored in the context of ion Coulomb crystals[121] and magnetic fields coupled to quantum many-body systems.[122] Quantum solitons are expected to exhibit long coherence times in the presence of cooling in the Doppler limit, and can be manipulated thanks to the spectral properties of the internal modes, which have been proposed as carriers of quantum information.[123] As a test-bed for entanglement generation[124] and the subsequent decoherence,[125] the creation of quantum structural defects might shed new light on fundamental issues concerning the relation between decoherence and critical phenomena.[126]

11. Soliton Formation in Bose–Einstein Condensation

One of us has suggested the finite-rate Bose–Einstein condensation of a thermal cloud in an elongated trap as an inhomogeneous test-bed for the KZM.[93] The inhomogeneity of the trap plays the key role in re-setting the dependence between the quench rate and the number of defects — solitons in a BEC "cigar" (see Fig. 9). The study of Bose–Einstein condensation in a harmonic trap motivated the development of the IKZM theory we have presented in Sec. 9.

This proposal has recently been realized in the laboratory at the BEC center in Trento.[16] The basic idea is that as a thermal cloud of atomic vapor undergoes evaporative cooling through the critical temperature for Bose–Einstein condensation, different regions of the newborn condensate pick up a different condensate phase. When two neighboring regions merge, the mismatch in the phase of the condensate wavefunction acts as a seed for the formation of a phase jump and the corresponding density dip: a

64

gray soliton is spontaneously formed. Numerical simulations based on the stochastic Gross–Pitaevskii equation indicate that this scenario in a homogeneous cloud is well-described by the Kibble–Zurek mechanism.[100,127] As an instance of a single realization, Fig. 13 shows the time evolution of the density profile of a newborn condensate following an evaporative cooling ramp. From the trajectory of the density dips, it is apparent that these excitations constitute spontaneously formed solitons. Note however that in thermal equilibrium solitons are as well expected to be formed.[128]

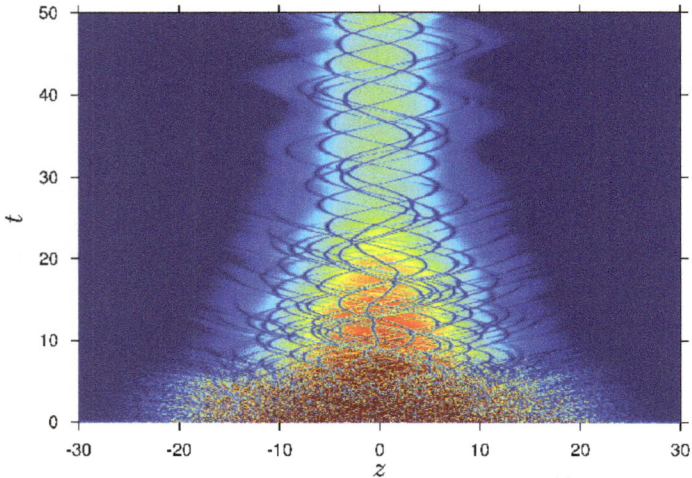

Fig. 13.   Time evolution of the density profile in a single realization of the cooling ramp simulated with the classical field method.[100] The initial state is chosen from a canonical ensemble above the critical temperature for Bose–Einstein condensation. Evaporative cooling is simulated by a linear ramp of the axial trapping potential using a one-dimensional generalized Gross–Pitaeveskii equation. Courtesy of E. Witkowska and P. Deuar.

In harmonic traps, the dynamics of Bose–Einstein condensation is more complex due to the inhomogeneous nature of the system. The atomic cloud is trapped in an anisotropic three dimensional harmonic confinement $U(r, z)$ with a cigar-shape, characterized by an axial frequency ω and a transverse one $\omega_\perp$ ($\omega_\perp > \omega$). The density of the cloud is highest at the center of the trap. Disregarding the transverse degrees of freedom, one can use the local density approximation to estimate the value of the critical temperature based on the Einstein equation with an axial dependence,

$$T_c(z) = \frac{2\pi\hbar^2}{mk_B}\left[\frac{\rho(r=0,z)}{\zeta(3/2)}\right]^{\frac{2}{3}}, \qquad (59)$$

which is obtained by replacing the constant density ρ by $\rho(r=0,z)$ in the expression for the critical temperature in a homogeneous system.

In experiments (e.g. at Trento[16]), a radio-frequency knife is used to force the evaporation of the cloud, by flipping the atomic spin from a trapped state to an untrapped state. Atoms with a potential energy $U_{RF} = h\nu_{RF}$ measured with respect to the bottom of the trap are forced to evaporate. The resulting axial temperature profile is given by

$$T(z) = \frac{U_{RF} - U(r,0,z)}{\eta k_B}. \qquad (60)$$

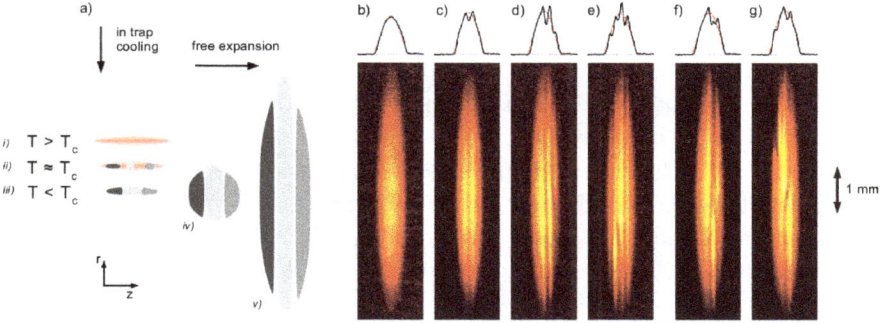

Fig. 14. Spontaneous soliton formation under forced evaporative cooling of a cigar-shaped atomic cloud.[16] (a) As the temperature (i)–(iii) is decreased below its critical value for Bose–Einstein condensation, causally disconnected regions of the newborn condensate cloud pick up different phases of the condensate wavefunction, and the subsequent dynamics leads to the formation of solitons. Under time of flight (iv)–(v), an initial cigar-shaped cloud expands mainly along the transverse direction. (b)–(g) Snapshots of the density profile after time of flight of clouds containing 0, 1, 2, 3, solitons and two instances exhibiting bending of the soliton in the transverse direction. Fits to the self-similar expansion of the Thomas–Fermi density profile (red line) are compared with integrated density profiles of the central region of the cloud (black line). Reprinted by permission from Macmillan Publishers Ltd: Nature Physics © (2013), Ref. 16.

In-situ optical imaging of solitons is challenging due to the smallness of the typical values of the healing length, which sets the width of the soliton. As a result, experimental measurements often resort to time-of-flight (TOF) imaging which magnifies the size of the cloud, see Fig. 14. When interactions can be disregarded during TOF, the dynamics is essentially ballistic, and the evolution of local correlations such as the density profile is rather trivial. This is the case when the anisotropy of the trap is not too large. When the confined cloud acquires an effectively one dimensional character, no true BEC is possible,[129] and the presence of phase fluctuations in the trapped superfluid severely distorts the TOF dynamics. Counting of solitons by imaging the density profile of the cloud after time of flight is then hindered by the fact that phase fluctuations map to density fluctuations as suggested in Ref. 130, and experimentally demonstrated in Ref. 131. This mapping precludes establishing a correspondence between the density of fringes in the expanded cloud and the initial number of solitons, formed spontaneously as a result of evaporative cooling, presumably described by the IKZM. As a consequence, the anisotropy should be low enough to minimize phase fluctuations associated with the low-dimensional character of the cloud. At the same time, it should be high enough to reduce the role of the instability and decay mechanisms of solitons in elongated three dimensional clouds, such as the snake instability. This is the regime of relevance to the experiment.[16] Indeed, TOF snapshots (f) and (g) in Fig. 14 suggest bending of the soliton along the transverse degree of freedom in the trapped cloud. The time of flight achieved to image the system were particularly large since TOF was assisted by levitation. We point out that as an alternative, shortcuts to adiabatic expansions[55,132,133] can be used with similar outcomes.[134,135]

Table 2. Power laws predicted by the Kibble–Zurek mechanism for the number of solitons N_s as a function of the cooling rate $1/\tau_Q$ induced by forced evaporation through the critical temperature for Bose–Einstein condensation of a cigar-shaped atomic cloud. The exponent α of the power-law $N_s \sim \tau_Q^{-\alpha}$ is shown for different critical exponents (ν, z) and trapping potentials.

Critical exponents	Homogeneous system	Harmonic trap
Arbitrary (ν, z)	$\dfrac{\nu}{1+\nu z}$	$\dfrac{1+2\nu}{1+\nu z}$
Mean-field theory $(\nu = \frac{1}{2}, z = 2)$	$\frac{1}{4}$	1
Experiments/F model[137] $(\nu = \frac{2}{3}, z = \frac{3}{2})$	$\frac{1}{3}$	$\frac{7}{6}$

The upshot of the counting statistics was a power-law dependence of the mean number of solitons N_s on the cooling rate $1/\tau_Q$ induced by forced evaporation, i.e. $N_s \sim \tau_Q^{-\alpha}$. The power-law exponent resulting from a fit to the experimental data was found to be $\alpha = 1.38 \pm 0.06$, which clearly deviates from the homogeneous KZM exponent and suggests a possible agreement with the IKZM, as illustrated in Table 2. This exponent was shown to be robust against variations of the mean atom number of the newborn condensate (which was varied from 4 to 25 million atoms), an indication of universal behavior.

The KZM as well as the IKZM can be used as a tool to determine critical exponents,[127] an application of interest in Bose–Einstein condensation. The experiment[136] reported deviations from mean-field behavior, $\nu = 1/2$. The BEC transition is believed to belong to the static 3D XY universality class, for which $\nu = 0.6717(1)$ according to the theoretical estimate in Ref. 91. Let us ignore for the moment possible systematic experimental errors, and assume that the power-law exponent α measured in the Trento experiment is actually given by the IKZM, $\alpha = \frac{1+2\nu}{1+\nu z}$. In principle, one can extract the value of the dynamical critical exponent $z \simeq 1.04 \pm 0.11$, which has not been directly measured so far in experiments. This would rule out both the mean-field value $z = 2$ (by 8-σ) as well as the F model $z = \frac{3}{2}$ value[137] (by 4-σ), when the experimental data is taken "at the face value", i.e. without accounting for the possible systematic effects, such as decay of solitons (that may steepen the dependence) as well as the fact that the number of solitons created in the trap is of order 1 (which may result in steepening of the dependence of their number on the quench rate we have discussed in the previous section). There is also a concern signalled by the authors of the experiment that the axial temperature in the BEC cloud is not uniform, which may further modify the scaling. Experiments in larger traps that lead to more solitons would be helpful in addressing this concern.

Numerical simulations using the classical field method[100] lead to similar power-law exponents which would agree with the IKZM, but are based on a model of evaporation along the axial direction which is not applicable to the experiments.

12. Vortex Formation in a Newborn Bose–Einstein Condensate

The observation of spontaneous soliton formation during Bose–Einstein condensation was actually preceded by analogous experiments in pancake-shaped atomic clouds.[10] The process is fairly similar. Consider a thermal quench between an initial temperature T_i and a final value T_f which is linear in time, i.e. $T(t) = T_i - t\frac{T_i - T_f}{\tau}$. During the cooling of the atomic cloud below the critical temperature for Bose–Einstein condensation, coherent patches are created where the phase of the condensate wavefunction is chosen independently and is approximately constant. When these different regions merge, there is a chance for the phase to accumulate along a close loop in (physical) space surrounding a given point. Indeed, it was shown experimentally that the explicit merging of three independent BEC clouds results in the formation of a vortex with a certain probability given by the geodesic rule.[138] Consider the expectation value of the order parameter $\langle \hat{\psi} \rangle_0 = |\psi| e^{i\theta(x)}$. The phase accumulated around a loop should be an integer modulo 2π and can be characterized by the winding number $\mathcal{W} = \frac{1}{2\pi} \oint \partial \theta$. The fundamental (or first) homotopy group is indeed given by the ring of integers, $\pi_1(U(1)/\{1\}) = \pi_1(S^1) = \mathbb{Z}$ (see Appendix A). Whenever $|\mathcal{W}| \geq 1$ a line defect, string or vortex is formed. In a homogeneous system, or for fast enough quenches in a trapped cloud, the density of vortices is expected to be given by Eq. (10), i.e. the homogenous KZM scaling,

$$ n = \frac{1}{f^2 \hat{\xi}^2} = \frac{1}{f^2 \xi_0^2} \left(\frac{\tau_0 2\delta}{\tau T_c(0)} \right)^{\frac{2\nu}{1+\nu z}} , \tag{61} $$

while when the influence of the harmonic confinement is taken into account[93,99] (now in an approximately 2D BEC "pancake", rather than a "cigar"),

$$ n \propto \left(\frac{\tau_0 2\delta}{\tau T_c(0)} \right)^{\frac{2(1+2\nu)}{1+\nu z}} , \tag{62} $$

as discussed in Sec. 9. The experiment[10] also reported the spontaneous vortex formation in a toroidal trap. This geometry offers the opportunity to explore a scenario where the condensation is inhomogeneous in the transverse degree of freedom and homogeneous in the toroidal direction. The density of vortices is predicted to scale then as[99]

$$ n \propto \left(\frac{\tau_0 2\delta}{\tau T_c(0)} \right)^{\frac{1+3\nu}{1+\nu z}} . \tag{63} $$

The accurate experimental determination of the power-law exponents in any pair of these three scenarios, which remain untested to-date, would allow the independent determination of the critical exponents ν and z.

13. Mott Insulator to Superfluid Transition

A natural testing ground for the KZM in quantum phase transitions is the transition between a Mott insulator (MI) and a superfluid (SF) phase, exhibited by the Bose–Hubbard model[139,140] and accessible to quantum simulation based on ultracold gases in optical lattices.[141,142] The Hamiltonian of the system is

$$\hat{\mathcal{H}}_{BH} = -\sum_{\langle i,j\rangle} J_{ij}(\hat{b}_i^\dagger \hat{b}_j + h.c.) + \frac{U}{2}\sum_i \hat{n}_i(\hat{n}_i - 1) - \mu \sum_i \hat{n}_i \quad (64)$$

where μ is the chemical potential, the tunneling matrix element for an atom to hop from site i to j is given by J_{ij}, $\hat{b}_i$ and $\hat{b}_i^\dagger$ are respectively the annihilation and creation operators, and U is the on-site interaction. We shall simplify the discussion by setting $J_{ij} = J$. In actual experiments the optical lattice implementing $\hat{\mathcal{H}}_{BH}$ is usually contained in a harmonic trap that will result in, e.g. the couplings J_{ij} that are spatially dependent.

In the limit $J/U \ll 1$ the homogeneous system is in the Mott insulator phase whenever the filling is commensurate with n atoms per site. The many-body state takes the form $|MI\rangle \propto \prod_{i=1}^M (\hat{b}_i^\dagger)^n |0\rangle$. The opposite limit $J/U \gg 1$ corresponds to the superfluid phase with a many-body state of the form $|SF\rangle \propto \left(\sum_{i=1}^M \hat{b}_i^\dagger\right)^N |0\rangle$, characterized by phase coherence and large fluctuations in the number of particles per site. The transition from the MI to the SF can be driven by increasing the relative coupling $(J/U)(t)$ in a finite time τ_Q.

The buildup of correlations induced by such a quench has been analyzed in a series of works.[27,143–147] In a 3D experimental realization of the Bose–Hubbard model, an interesting systematic study of the amount of excitations and energy produced during a quench as a function of the quench rate was undertaken, and a power-law dependence was found suggesting a KZM-like behavior.[148] The amount of excitations was estimated by comparing the density profile after time of flight (TOF) for an initial cloud in the ground state ($n_0(x,y;t)$) and an out-of-equilibrium cloud prepared by a quench ($n(x,y;t)$). The following quantity was used

$$\tilde{\chi}^2 = c\mathcal{N} \int dxdy \frac{[n(x,y;t) - n_0(x,y;t)]^2}{n_0(x,y;t)} \quad (65)$$

where $\mathcal{N} = \int dxdy n(x,y;t)$ and c is a constant chosen in agreement with numerical simulations. The dependence of $\tilde{\chi}^2$ was fitted to a power-law $\tilde{\chi}^2 \propto \tau_Q^{-\alpha}$ with $\alpha = 0.31 \pm 0.03$. The kinetic energy $K = m\langle r^2\rangle/(2t_{TOF})$, where t_{TOF} is the time of flight and $\langle r^2\rangle = N\int dxdy(x^2 + y^2)n(x,y;t)$,

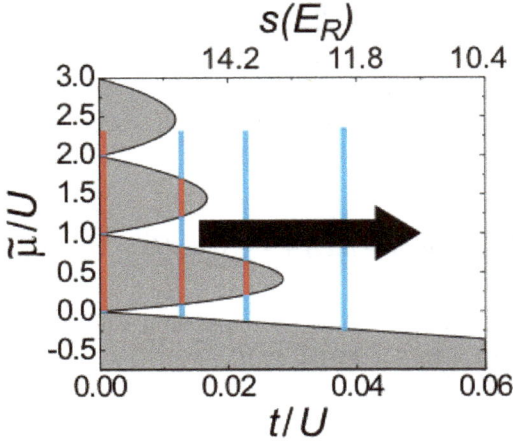

Fig. 15.   Phase diagram of the Bose–Hubbard model illustrate the boundary between the Mott insulator and superfluid phases. The vertical lines range over the densities and effective chemical potential $\tilde{\mu}$ sampled in a single experimental realization. The quench is driven by a fast modulation of $J = t$ in the direction of the black arrow. From Chen *et al.*[148] © 2011 American Physical Society.

was also analyzed and fitted to a power law with a comparable exponent $\alpha = 0.32 \pm 0.02$.

According to Chen *et al.*, these exponents deviate from theoretical predictions for a 3D homogeneous transition. They studied the dependence on the quench rate of the density fluctuations characterized by $\tilde{\chi}^2$ and the kinetic energy, and compared it with a power-law with exponent $\alpha = \frac{3\nu}{\nu z+1}$, i.e., which would result when the fluctuations and excess of kinetic energy are the same in each domain of size $\hat{\xi}$. For $\nu = 1/2$ and $z = 2$ this would lead to $\alpha = 3/4$.[148]

We shall not attempt to justify these assertions here, nor shall dispute them. Clearly, if one were to accept validity of the homogeneous KZM for this relatively modestly sized inhomogeneous system of about 1.6×10^5 atoms (a BEC-filled optical lattice confined to an inhomogeneous sphere of radius of only ~10 sites, each occupied by ~ 3 atoms), more detailed analysis would be useful to explore the conjecture that the applied measures of the degree of excitation imparted by the quench result in the behavior described simply by the same kinetic energy and same departure from homogeneity in each ~ $\hat{\xi}^{-3}$ volume, which is what the ansatz[148] described above suggests. Moreover, the experimental conditions hindered a direct connection with this KZM estimate. In particular, the inhomogeneous character

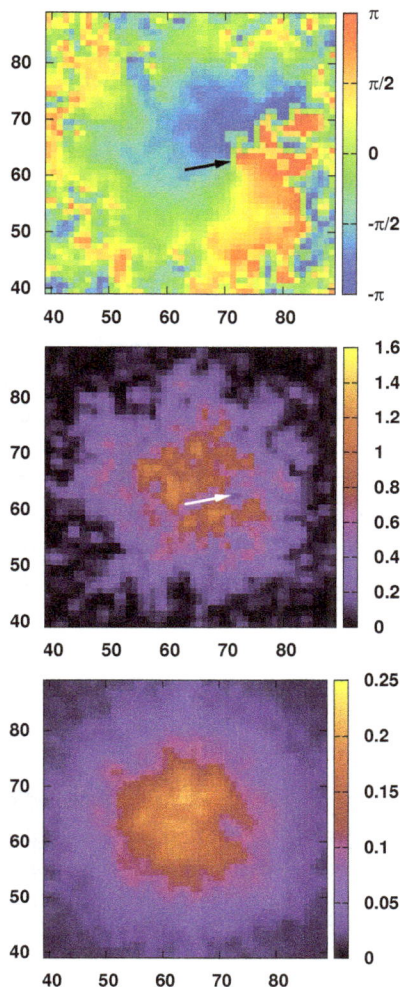

Fig. 16.   Simulation of a linear quench from the Mott insulator to superfluid phase in a three dimensional trapped atomic cloud using a discrete Gross–Pitaevskii equation. The top and medium panels correspond to phase and density in a x–y cross section across the center of the trap. The integrated density along the perpendicular direction is shown in the lower panel. The position of a vortex is indicated by an arrow. As noted in the discussion, the scaling of, e.g. the kinetic energy with the quench rate is similar to what was observed on the experiment by Chen *et al.* This coincidence between the effectively classical simulation (attained in the large occupancy per site limit) and the experiment (that had about 3 atoms per site) may be accidental, or may be due to the fact that BEC in the experiment was decohered[149] by, e.g. the finite temperature effects and the significant normal fraction. From Dziarmaga *et al.*[146] © 2012 American Physical Society.

of the system induced by the presence of an external harmonic trap results in an initial state with Mott insulating layers of different filling factors being separated by (presumably phase coherent) superfluid layers, see Fig. 15. In addition, the phase boundary is crossed at a range of densities, with different MI layers (corresponding to different lobes in the phase diagram) crossing the phase transition at different times. The analysis of this scenario is substantially more complicated than the exposition of the IKZM discussed in Sec. 9, where power-law behaviors are expected for a single critical front. The finite-temperature of the initial cloud in Ref. 148 has also not been taken into account in KZM studies applied to this transition.

We note that the numerical simulations[146] in the limit of large occupancy per site yield power laws, e.g. for the kinetic energy, that are similar to those observed in the experiment. Moreover, topological defects appear in the superfluid left after the Mott insulator — superfluid phase transition, see Fig. 16. It is again far from clear whether this coincidence of scalings is significant. In the large occupancy regime the system is effectively classical (which is what makes the computer simulation possible in the first place). On the other hand, in the actual experiment there was a substantial ($\sim 10\%$) normal fraction and nonnegligible temperature. That combination may cause decoherence[149] and, hence, force a quantum many-body system to behave in an effectively classical manner.

14. Summary and Outlook

The Kibble–Zurek mechanism reviewed here is based on the combination of two key ideas. The seminal observation of Tom Kibble[1,2] made it clear that, at least in the cosmological context, phase transitions expected to occur as the Universe cools soon after the Big Bang will result in a mosaic of domains of the size close to the Hubble radius at the time of the transition. This is simply a consequence of relativistic causality — domains are forced to break symmetry independently, and, hence, at random. Moreover, when the resulting homotopy group is nontrivial, phase ordering cannot completely smooth out the post-transition configurations of the order parameter, as the random choices of broken symmetry lead to irreconcilable differences that crystalize into topological defects.

In the second order phase transitions encountered in the laboratory relativistic causality does not yield useful limits, but the speed of light can be effectively replaced by the relevant speed of sound[19,20,23] leading to an estimate of the size $\hat{\xi}$ of the domains that can consult on how to break the symmetry, and, hence, that can choose to break symmetry more or less in unison. The resulting density of topological defects and other excitations, left behind by phase transitions induced at a finite speed, depends on the interplay of the quench rate (the rate at which the critical point is traversed) and critical slowing down (the rate with which systems can adjust), and as a result, on the universality class of the transition. The scaling of $\hat{\xi}$ with the quench rate (reflected in the density of defects left behind by the transition) can be investigated in the laboratory.

Experiments testing KZM scaling were the focus of our review. The Kibble–Zurek scaling was also tested in classical and quantum phase transitions in a variety of computer experiments[22,31,86,100,112,113,127,150–166] and analytical works,[46,60,61,167–170] and found to hold essentially whenever it was expected to apply. Laboratory experiments are, of course, more difficult. Above all, it is hard to vary the quench rate over several orders of magnitude (needed to detect the fractional power laws predicted for $\hat{\xi}$ as a function of τ_Q) while avoiding effects that can either suppress generation of topological defects (e.g. inhomogeneities) or result in formation of defects in processes (e.g. convection in superfluids) independent of KZM that could obscure KZM-predicted scaling. Moreover, defects formed in the course of the transition can annihilate during the phase ordering that follows the transition.

A brief summary of the present day "experimental KZM landscape" is that there are now several experiments that have found, in various systems, results consistent with KZM scalings. However, all of them require at present caveats and additional assumptions for interpretation.

Switching between nonequilibrium steady states provided early evidence for KZM scaling.[73] Nonetheless, subsequent experiments as well as numerics indicated that in such situations where the renormalization theory cannot be invoked KZM scalings may be only an approximation or do not apply.[79] Still, such efforts have led to the earliest experimental indications of the KZM scaling, and may offer intriguing opportunities for extension of KZM to transitions that are not described by renormalization or even by partial differential equations.

Trapping of flux quanta in tunnel Josephson junctions yielded scaling that appears to be reliable, but the detected exponent of ~ 0.5 was twice what was initially expected. That expectation was based on the prediction of the doubling of the power law for large winding numbers.[85] Recent analysis[87] of the winding numbers in the case of small loops ($\mathcal{C} \ll \hat{\xi}$) indicates that, while one would indeed expect the exponent that governs the *dispersion* of $\mathcal{W}$ to double in the regime where $|\mathcal{W}| > 1$ is vanishingly unlikely, the *frequency* of trapping of $|\mathcal{W}| = 1$ scales with four times the power predicted for large $|\mathcal{W}|$. This suggests that the KZM accounts for the experimental results. This quadrupling may be also relevant for small superconducting loops, where the observed frequency of trapping a flux quantum scales with the exponent $\simeq 0.62 \pm 0.15$,[86] consistent with 0.5 seen in tunnel Josephson junctions[84] (although possibly suggestive of $\frac{2}{3}$, which is — one might be even tempted to speculate — four times $\frac{1}{6}$, the exponent expected for the scaling of typical winding numbers trapped by large loops for $\nu = \frac{1}{2}$ and $z = 1$ or a superfluid with $\nu = \frac{2}{3}$ and $z = \frac{3}{2}$, where $\hat{\xi} \sim \tau_Q^{\frac{1}{3}}$).

We note that all these discussions of doubling and quadrupling ignore the role of the magnetic field, which, as was pointed out in the case of loops[23] and demonstrated much more clearly in 2D systems with the help of numerical simulations[152–154] may play a significant role in flux trapping and defect formation in systems with local gauge invariance.

Defect formation in multiferroics is a new frontier. Experiment in $ErMnO_3$ yields a compelling power law,[89] but its interpretation in terms of KZM depends on the nature of the critical region of the transition that is inaccessible to, e.g., susceptibility measurements, as a result of the high critical temperature. Still, theoretical analysis[90] based on the 3D XY model yields an impressive agreement of KZM with the experiments. Nevertheless, a more precise determination of the exponent that governs the power law scaling that would clearly establish the connection with the ν and z predicted for the 3D XY universality class would be welcome: it would amount to the first experimental confirmation of KZM scaling in a setting that is not mean field.

Generalization of the KZM to inhomogeneous systems is usually needed to interpret experiments in harmonically trapped ions and BECs. Formation of kinks in ion Coulomb crystals[13–15] and solitons in Bose–Einstein condensation[16] has been recently reported. It has been argued in both ion crystals and BECs that the data are consistent with the KZM when one recognizes both the consequences of inhomogeneity and (in the case of the ion chains) small size of the system. Dependence of the conclusions about scaling of these additional assumptions complicates the interpretation especially in the case of kinks in ion chains, but the results are consistent with the suitably modified versions of the KZM.

In the case of BEC solitons the measured power laws are close to the analytic predictions,[93] and the remaining discrepancy may be due to the difference between the simplified effect of the harmonic trap analyzed theoretically[93] and the experimental reality.[16] Indeed, corrections to a power-law scaling can be expected in inhomogeneous systems.[92,99] Additional experimental results and theoretical as well as numerical efforts would certainly be useful.

The presence of vortices in a newborn BEC has also been detected.[10,138] They have appeared presumably as the result of the KZM. However, obtaining reliable power laws in this case is even harder. The experiments are carried out, e.g. in approximately 2D BEC pancakes, so inhomogeneities and small systems sizes will play a role and complicate the analysis.

An experiment, that has not been completed as yet but is under way by the group of Markus Oberthaler,[171] probes a quantum miscibility-immiscibility phase transition in a Bose–Einstein condensate.[172] Numerical simulations in an effectively 1D toroidal trap yield scalings of the size of domains of the two hyperfine BEC states that are in good agreement with the KZM prediction.[159,160] In the harmonic trap (where the actual experiments will likely take place) inhomogeneity modifies domain sizes in a way that influences the observed power law, again complicating direct comparisons with the KZM predictions, although numerical simulations may help.

Experimental investigations of the quantum KZM (exemplified by the miscibility-immiscibility transition) are only the beginning. Experiments to date (e.g. related to the Bose–Hubbard model[148]) suffer from complications caused by the inhomogeneities and small system sizes. In view of the rather complicated phase diagram of the Bose–Hubbard model, inhomogeneities make critical exponents relevant for the KZM scaling difficult to infer. Moreover, computer simulation of the quantum Bose–Hubbard model are difficult, as systems of sizes large enough to hope for a suitably well defined quantum phase transition are also large enough to be essentially out of reach of present day computers. One can study larger systems only in the limit where they become effectively classical — when the number N of atoms per site is large.[146,147] When one compares results of such simulations with the data obtained in experiments, there are no obvious discrepancies that cannot be blamed on inhomogeneity or finite size, but this rather tentative conclusion (based on the comparison of a classical simulation to a quantum many-body system) is unsatisfying and it certainly leaves plenty of room for improvement.

This cautious assessment of the present status of the experiments on the dynamics of quantum phase transitions is likely to be revised in the near future. Moreover, quantum phase transitions in the Ising model (which is much better understood theoretically) may be eventually implemented (e.g. by emulating its dynamics[173,174]) in suitably large systems. This would be interesting not just because of the implications for the KZM, but because one could then study nonlocal superpositions of topological defects — "topological Schrödinger cats" or "Schrödinger kinks"[120] — as well as probe the possibility of a quantum superposition of distinct phases of matter.[122] Last, but not least, there are examples (e.g. quantum Ising model) where the KZM predicts the range of entanglement in the post-transition state[124] and even the effectiveness of the system undergoing second order phase transition as an environment responsible for decoherence.[126]

Kibble–Zurek mechanism employs *equilibrium* behavior of the system to predict nonequilibrium consequences of the dynamics of symmetry breaking. When we compare the first experimental tests of the KZM with some of the recent experiments, one important difference stands out: in the pioneering tests of the KZM the equilibrium behavior of the systems in the vicinity of the critical point was generally very well known as a result of earlier measurements, so the scaling predicted by the universality class was beyond doubt. This is often not the case in the recent KZM-inspired experiments of, for example, quantum phase transitions in optical lattices. It would seem prudent to test the equilibrium of the actual system as a prerequisite, and to verify that the scalings predicted by the universality class indeed capture its equilibrium behavior, or, at the very least, to evaluate the extent and nature of the departures before embarking on tests of the KZM. Of course, there is usually a microscopic theory (e.g. Bose–Hubbard), but its implication for the critical regions are typically well-established only for an infinite homogeneous system, and the extent to which it is a good approximation of an often modestly sized and inhomogeneous system available in the laboratory is frequently not known. Moreover, it is often far from clear how to apply that theory to what is measured in the experiment (e.g. critical exponents may differ depending on how the critical region is traversed in the Mott insulator-superfluid transition[139]).

To sum up, we note that the already considerable progress in verifying the KZM achieved in this millenium has accelerated in the past few years. Given the broad applicability of the KZM, it seems likely that the study of phase transition dynamics will remain an exciting research field in the foreseeable future. Our focus on experiments involving the scaling of topological defects is understandable, given the roots of the KZM. There are however other excitations of the order parameter that may be left in far-from-equilibrium state due to the KZM, and that can be used to test it. We have discussed solitons and vortices in BEC as examples, but even more transient excitations (e.g. those created in superconductors[175,176] or left behind by the chiral symmetry breaking in ^{3}He[177]) may be of interest in this respect.

Acknowledgments

We thank Tom Kibble for the insightful contributions that have started the field we have attempted to review, and for many useful, pleasant, and inspiring interactions. It is also a pleasure to thank Sang-Wook Cheong, Franco Dalfovo, Bogdan Damski, Brian DeMarco, Piotr Deuar, Jacek Dziarmaga, Gabriele Ferrari, Uwe Fischer, Jerome Gaunlett, Guo-Ping Guo, Wenceslao González-Viñas, Paul Haljan, Valery Kiryukhin, Giacomo Lamporesi, Chuan-Feng Li, Shi-Zeng Lin, Tanja E. Mehlstaubler, Emilie Passemar, Marek M. Rams, Ray J. Rivers, Kazimierz Rzazewski, Nikolai Sinitsyn, Grigori E. Volovik, Emilia Witkowska and Vivien Zapf for useful comments and discussions.

This work is supported by the U.S Department of Energy through the LANL/LDRD Program and a LANL J. Robert Oppenheimer fellowship (AD).

Appendix A.   Spontaneous Symmetry Breaking: the Role of Topology

Spontaneous symmetry breaking arises in situations where a symmetry of the system is not manifested in its ground state, and it is a phenomenon tied to the degeneracy of the later.[1,25] A well-known example is the breakdown of rotational invariance in a ferromagnet. This scenario is relevant in cosmology,[1,2] elementary particle physics,[178,179] and condensed matter systems.[23,25] We next summarize the basics of homotopy theory and its use in this context, at a somewhat technical level. Consider the case in which the Hamiltonian (or free-energy functional) $\hat{\mathcal{H}}$ of the system is invariant under an operation g of the symmetry group G, which is represented by a unitary transformation $U(g)$,

$$U^{-1}(g)\hat{\mathcal{H}}U(g) = \hat{\mathcal{H}}, \forall g \in G. \tag{A.1}$$

Now, assume that there exists an order parameter, and operator $\hat{\psi}$ whose ground-state expectation value is not invariant under G, i.e.

$$\langle 0|U^{-1}(g)\hat{\psi}U(g)|0\rangle = D\langle\hat{\psi}\rangle_0 \neq \langle\hat{\psi}\rangle_0, \tag{A.2}$$

where D is a rotation matrix. That is, the states $U(g)|0\rangle$ and $|0\rangle$ are nonequivalent, but are degenerate according to (A.1). Typically, different phases of the system will have a symmetry group, a subgroup of G called the isotropy group H, which represents the leftover symmetry in the broken-symmetry phase. An arbitrary element $h \in H$ leaves invariant the order parameter $\hat{\psi}$, $h\hat{\psi} = \hat{\psi}$. The order parameter manifold $\mathcal{M}$ of degenerate vacuum states is homeomorphic (from the Greek for "similar shape", a relation denoted by the symbol $\simeq$) to the (left) coset space of H in G,[180]

$$M \simeq G/H. \tag{A.3}$$

The simplest example is that of the linear to zigzag transition that for homogeneous ion chains is characterized by $G = \mathbb{Z}_2$, $H = e$ (e being the identity $\{1\}$), and $\mathcal{M} \simeq \mathbb{Z}_2$, discussed in detail in Sec. 10. Sections 7, 11, and 12 are devoted to the BEC transition associated with a scalar order parameter, where $G = U(1)$, $H = \{1\}$, and $G/H = U(1)$. Symmetry breaking in spinor Bose–Einstein condensates is more complex, and its characterization has recently led to a large body of research.[181–183] following the observation of spin textures in the laboratory.[9] For instance, a spin-1 BEC is characterized by a $G = U(1) \times SO(3)$ resulting from the invariance under $U(1)$ gauge transformations and rotations in spin space, and that can lead to a variety of symmetry breaking scenarios.[181]

Homotopy theory deals with continuous transformations between objects that belong to the same equivalence class and it can be used for the systematic classification of topological excitations. Let $I = [0, 1]$ and consider two continuous maps $f, g : X \to Y$ between topological spaces X and Y. A homotopy between f and g is a continuous map $F : X \times I \to Y$ satisfying $F(x, 0) = f(x)$, $F(x, 1) = g(x)$, $\forall x \in X$. Provided F exists, f is said to be homotopic to g, which is symbolically denoted by $f \sim g$. This is an equivalence relation satisfying reflectivity ($f \sim f$), symmetry ($f \sim g$ implies $g \sim f$), and transitivity (if $f \sim g$ and $g \sim h$ then $f \sim h$). A path with initial point x_0 and final point x_1 is a continuous map $\alpha : I \to X$ such that $\alpha(0) = x_0$ and $\alpha(1) = x_1$. A path for which $x_0 = x_1$ is called a loop with base point x_0, this is, a loop in which the boundary ∂I of $I = [0, 1]$ is mapped to x_0. We shall refer to two specific types of loops below, a constant loop $c : I \to X$ which has a fixed image in X $\forall t \in I$, and an inverse loop $\alpha^{-1}(t) \equiv \alpha(1 - t)$ $\forall t \in I$. Two loops $\alpha, \beta : I \to X$ with base point x_0 are homotopic ($\alpha \sim \beta$) given that an homotopy $F : I \times I \to X$ exists, i.e. a continuous map $F : I \times I \to X$ can be found that satisfies $F(t, 0) = \alpha(t)$, $F(t, 1) = \beta(t)$ $\forall t \in I$ and $F[0, t'] = F[1, t'] = x_0$ $\forall t' \in I$. The set of all loops with base point x_0 can be classified into homotopy classes.

A homotopy class $[\alpha]$ is the set of loops which are homotopic to α. The fundamental group or first homotopy group is the set of all homotopy classes of loops with base point x_0. It is denoted by $\pi_1(X, x_0)$ and satisfies the group properties with respect to the product of homotopy classes. This product is defined by $[\alpha] \cdot [\beta] = [\alpha \cdot \beta]$, where $\alpha \cdot \beta$ is the product of loops α and β in which α is first traversed and then β is traversed. Specifically, the product of homotopy classes in $\pi_1(X, x_0)$ satisfies

$$([\alpha] \cdot [\beta]) \cdot [\gamma] = [\alpha] \cdot ([\beta] \cdot [\gamma] \tag{A.4}$$

$$[\alpha] \cdot [c] = [c] \cdot [\alpha] = [\alpha] \tag{A.5}$$

$$[\alpha] \cdot [\alpha^{-1}] = [\alpha^{-1}] \cdot [\alpha] = [c] \tag{A.6}$$

where the identity element $[c]$ is given by the set of loops homotopic to a constant loop.

There exists an isomorphism (a bijective homomorphism) between fundamental groups $\pi_1(X, x_0)$ and $\pi_1(X, x_1)$ of loops within the same connected topological spaces X with different base points x_0 and x_1 which allows us to use the simplified notation $\pi_1(X)$ for the fundamental group. A mapping from a loop to the unit circle S^1 is described by the isomorphism between $\pi_1(S^1)$ and $\mathbb{Z}$, where the integer winding number corresponds to the number of times the loop wraps around the unit circle. Higher homotopy groups are defined in a similar way to π_1 by considering homotopy classes of the n-sphere $S^n = \{x \in \mathbb{R}^{n+1} | |x|^2 = 1\}$. Let us consider the n-cube $I^n = I \times \cdots \times I = \{(s_1, \ldots, s_n) | s_i \in [0, 1] \ \forall 0 \le i \le n\}$ with boundary $\partial I^n = \{(s_1, \ldots, s_n) \in I^n | s_i = 0 \text{ or } 1\}$. A map $\alpha : I^n \to X$ that maps the boundary ∂I^n to a point x_0 is a n-loop. When a homotopy exists between n-loops α and β, they are said to be homotopic, and the set of n-loops homotopic to a given n-loop α constitutes a homotopy class $[\alpha]$. The nth homotopy group of n-loops with base point x_0 is given by the set of homotopy classes of n-loops.

The classification of topological excitations is achieved by the homotopy groups $\pi_n(\mathcal{M})$ of the order parameter manifold $\mathcal{M}$ with the dimension of homotopy being given by $n = D - d - 1$ in terms of the spatial dimension D and the dimension of the (singular) topological excitation d (for nonsingular topological excitations such as skyrmions, $n = D - d$). The homotopy groups $\pi_n(\mathcal{M})$ characterize mappings from the n-sphere S^n enclosing the topological excitation in real space into the vacuum manifold $\mathcal{M}$. Elements of a given group $\pi_n(\mathcal{M})$ belong to the same class of stable topological excitations, equivalent by continuous deformations. The number of *domains* or disconnected regions in $\mathcal{M}$ is given by $\pi_0(\mathcal{M})$ (formally π_0 lacks a group structure). If $\pi_0(\mathcal{M}) = k$, there are $k+1$ disconnected regions. When $\mathcal{M}$ is disconnected, topological excitations associated with the different choices of $\langle \hat{\psi} \rangle_0$ in space are known as *domain walls*, and are typically associated with the breakdown of a discrete symmetry, as in the linear to zigzag transition. One can next consider the change of the order parameter along closed loops in S^1 in real space within the same connected component of $\mathcal{M}$. If $\langle \hat{\psi} \rangle_0$ is a smooth function along the loop, then $\pi_1(\mathcal{M})$ is trivial and equal to the identity e. Otherwise, elements of the group $\pi_1(\mathcal{M}) \neq e$ characterize *line defects* or *strings*, such as quantized vortices in superfluids and scalar BEC, and flux tubes in type II superconductors, associated with $U(1)$ symmetry breaking. Topological excitations known as *monopoles* arise in the presence of noncontractive surfaces in $\mathcal{M}$ such as S^2, whenever $\pi_2(\mathcal{M}) \neq e$. They are associated with the breakdown of nonabelian symmetries to a subgroup containing $U(1)$. In $D = 3$, $d = 2$ for domains walls, $d = 1$ for strings, and $d = 0$ for monopoles. Three dimensional topological defects associated with nontrivial mappings from S^3 into $\mathcal{M}$ are characterized by the homotopy group $\pi_3(\mathcal{M})$ and are known as *textures* or nonsingular solitons.

Table A1.   Homotopy groups of certain vacuum manifolds.

$\mathcal{M}$	π_1	π_2	π_3
$U(1)$	$\mathbb{Z}$	0	0
$SU(n)$	0	0	$\mathbb{Z}$
$SO(3)$	$\mathbb{Z}_2$	0	$\mathbb{Z}$
S^2	0	$\mathbb{Z}$	$\mathbb{Z}$
S^3	0	0	$\mathbb{Z}$
S^4	0	0	0

The dynamics of symmetry breaking can in principle result in hybrid configurations with a variety of topological defects with different dimensions of homotopy and which can influence each other.[181,184] In that case, the classification in terms of π_n is no longer satisfactory, but Abe homotopy groups composed of possibly noncommutative groups π_1 and π_n can however account for topological excitations with $n \geq 2$. We refer the reader to Ref. 185 for a more detailed exposition and to Refs. 181–183 for a thorough discussion in the context of Bose–Einstein condensates. We close by pointing out that the use of conventional homotopy groups has limitations in the classification of topological defects located on the boundary of an ordered system, for which the use of relative homotopy groups has proven to be advantageous.[186,187]

References

1. T. W. B. Kibble, *J. Phys. A: Math. Gen.* **9**, 1387 (1976).
2. T. W. B. Kibble, *Phys. Rep.* **67**, 183 (1980).
3. I. Chuang, R. Durrer, N. Turok, B. Yurke, *Science* **251**, 1336 (1991).
4. M. J. Bowick, L. Chandar, E. A. Schiff, and A. M. Srivastava, *Science* **263**, 943 (1994).
5. V. M. H. Ruutu, V. B. Eltsov, A. J. Gill, T. W. B. Kibble, M. Krusius, Yu. G. Makhlin, B. Placais, G. E. Volovik, and W. Xu *Nature* **382**, 334 (1996).
6. C. Bäuerle, Yu. M. Bunkov, S. N. Fisher, H. Godfrin, and G. R. Pickett, *Nature* **382**, 332 (1996).
7. R. Carmi, E. Polturak, and G. Koren, *Phys. Rev. Lett.* **84**, 4966 (2000).
8. A. Maniv, E. Polturak, and G. Koren, *Phys. Rev. Lett.* **91**, 197001 (2003).
9. L. E. Sadler, J. M. Higbie, S. R. Leslie, M. Vengalattore, and D. M. Stamper-Kurn, *Nature* **443**, 312 (2006).
10. C. N. Weiler, T. W. Neely, D. R. Scherer, A. S. Bradley, M. J. Davis, and B. P. Anderson, *Nature* **455**, 948 (2008).
11. D. Golubchik, E. Polturak, and G. Koren, *Phys. Rev. Lett.* **104**, 247002 (2010).
12. M. Mielenz, H. Landa, J. Brox, S. Kahra, G. Leschhorn, M. Albert, B. Reznik, and T. Schaetz, *Phys. Rev. Lett.* **110**, 133004 (2013).
13. S. Ejtemaee and P. C. Haljan, *Phys. Rev. A* **87**, 051401(R) (2013).
14. S. Ulm, J. Roßnagel, G. Jacob, C. Degünther, S. T. Dawkins, U. G. Poschinger, R. Nigmatullin, A. Retzker, M. B. Plenio, F. Schmidt-Kaler, and K. Singer, *Nat. Commun.* **4**, 2290 (2013).
15. K. Pyka, J. Keller, H. L. Partner, R. Nigmatullin, T. Burgermeister, D. M. Meier, K. Kuhlmann, A. Retzker, M. B. Plenio, W. H. Zurek, A. del Campo, and T. E. Mehlstäubler, *Nat. Commun.* **4**, 2291 (2013).
16. G. Lamporesi, S. Donadello, S. Serafini, F. Dalfovo, and G. Ferrari, *Nature Phys.* **9**, 656 (2013).
17. P. C. Hendry, N. S. Lawson, R. A. M. Lee, P. V. E. McClintock, and C. D. H. Williams, *Nature* **368**, 315 (1994).
18. M. E. Dodd, P. C. Hendry, N. S. Lawson, P. V. E. McClintock, and C. D. H. Williams, *Phys. Rev. Lett.* **81**, 3703 (1998).
19. W. H. Zurek, *Nature (London)* **317**, 505 (1985).
20. W. H. Zurek, *Acta Phys. Pol. B* **24**, 1301 (1993).
21. T. W. B. Kibble and G. E. Volovik, *JETP Lett.* **65**, 102 (1997).
22. J. Dziarmaga, P. Laguna, and W. H. Zurek, *Phys. Rev. Lett.* **82**, 4749 (1999).

23. W. H. Zurek, *Phys. Rep.* **276**, 177 (1996).
24. W. H. Zurek, L. M. A. Bettencourt, J. Dziarmaga, and N. D. Antunes, *Acta Phys. Pol. B* **31**, 2937 (2000).
25. T. W. B. Kibble, in *Patterns of Symmetry Breaking* (Kluwer Academic Publishers, London, 2003).
26. T. W. B. Kibble, *Phys. Today* **60**, 47 (2007).
27. J. Dziarmaga, *Adv. Phys.* **59**, 1063 (2010).
28. A. Polkovnikov, K. Sengupta, A. Silva, and M. Vengalattore, *Rev. Mod. Phys.* **83**, 863 (2011).
29. A. Das, J. Sabbatini, and W. H. Zurek, *Sci. Rep.* **2**, 352 (2012).
30. J. Dziarmaga and W. H. Zurek, in preparation.
31. P. Laguna and W. H. Zurek, *Phys. Rev. Lett.* **78**, 2519 (1997).
32. A. Yates and W. H. Zurek, *Phys. Rev. Lett.* **80**, 5477–5480 (1998).
33. N. D. Antunes, L. M. A. Bettencourt, and W. H. Zurek, *Phys. Rev. Lett.* **82**, 2824 (1999).
34. N. D. Antunes, L. M. A. Bettencourt and W. H. Zurek. *Phys. Rev. D* **62**, 065005 (2000).
35. G. De Chiara, A. del Campo, G. Morigi, M. B. Plenio, and A. Retzker, *New J. Phys.* **12**, 115003 (2010).
36. L. Landau, *Phys. Soviet Union* **2**, 46 (1932).
37. C. Zener, *Proc. R. Soc. London A* **137**, 696 (1932).
38. E. C. G. Stueckelberg, *Helvetica Physica Acta* **5**, 369 (1932).
39. E. Majorana, *Nuovo Cimento* **9**, 43 (1932).
40. Yu. N. Demkov and V. I. Osherov, *Zh. Exp. Teor. Fiz.* **53**, 1589 (1967); *Sov. Phys. JETP* **26**, 916 (1968).
41. S. Brundobler and V. Elser, *J. Phys. A* **26**, 1211 (1993).
42. Yu. N. Demkov and V. N. Ostrovsky, *J. Phys. B* **28**, 403 (1995).
43. Yu. N. Demkov and V. N. Ostrovsky, *Phys. Rev. A* **61**, 032705 (2000).
44. V. N. Ostrovsky, *Phys. Rev. A* **68**, 012710 (2003).
45. N. A. Sinitsyn, *Phys. Rev. Lett.* **110**, 150603 (2013).
46. B. Damski, *Phys. Rev. Lett.* **95**, 035701 (2005).
47. B. Damski and W. H. Zurek, *Phys. Rev. A* **73**, 063405 (2006).
48. X.-Y. Xu, Y.-J. Han, K. Sun, J.-S. Xu, J.-S. Tang, C.-F. Li, and G.-C. Guo, arXiv:1301.2752 (2013).
49. C. Zhou, L. Wang, T. Tu, H.-O. Li, G.-C. Guo, H.-W. Jiang, and G.-P. Guo, arXiv:1301.5534 (2013).
50. M. Demirplak and S. A. Rice, *J. Phys. Chem. A* **107**, 9937 (2003); *J. Phys. Chem. B* **109**, 6838 (2005).
51. M. V. Berry, *J. Phys. A: Math. Theor.* **42**, 365303 (2009).
52. M. G. Bason, M. Viteau, N. Malossi, P. Huillery, E. Arimondo, D. Ciampini, R. Fazio, V. Giovannetti, R. Mannella, and O. Morsch, *Nature Phys.* **8**, 147 (2012).
53. J. Zhang, J. Hyun Shim, I. Niemeyer, T. Taniguchi, T. Teraji, H. Abe S. Onoda, T. Yamamoto, T. Ohshima, J. Isoya, and D. Suter, *Phys. Rev. Lett.* **110**, 240501 (2013).
54. M. Demirplak and S. A. Rice, *J. Chem. Phys.* **129**, 154111 (2008).

55. A. del Campo, *Phys. Rev. Lett.* **111**, 100502 (2013).

56. E. Torrontegui, S. Ibáñez, S. Martínez-Garaot, M. Modugno, A. del Campo, D. Guéry-Odelin, A. Ruschhaupt, X. Chen, and J. G. Muga, *Adv. At. Mol. Opt. Phys.* **62**, 117 (2013).

57. S. Sachdev, *Quantum Phase Transitions* (Cambridge University Press, Cambridge, 1999).

58. M. Lewenstein, A. Sanpera, V. Ahufinger, B. Damski, A. Sen De, and U. Sen, *Adv. Phys.* **56**, 243 (2007).

59. W. H. Zurek, U. Dorner, and P. Zoller, *Phys. Rev. Lett.* **95**, 105701 (2005).

60. J. Dziarmaga, *Phys. Rev. Lett.* **95**, 245701 (2005).

61. A. Polkovnikov, *Phys. Rev. B* **72**, 161201(R) (2005).

62. M. Kolodrubetz, B. K. Clark, and D. A. Huse, *Phys. Rev. Lett.* **109**, 015701 (2012).

63. R. Coldea, D. A. Tennant, E. M. Wheeler, E. Wawrzynska, D. Prabhakaran, M. Telling, K. Habicht, P. Smeibidl, and K. Kiefer, *Science* **327**, 177 (2010).

64. S. Korenblit, D. Kafri, W. C. Campbell, R. Islam, E. E. Edwards, Z.-X. Gong, G.-D.Lin, L.-M. Duan, J. Kim, K. Kim, and C. Monroe, *New J. Phys.* **14**, 095024 (2012).

65. A. del Campo, M. M. Rams, and W. H. Zurek, *Phys. Rev. Lett* **109**, 115703 (2012).

66. D.-H. Lee, G.-M. Zhang, and T. Xiang, *Phys. Rev. Lett.* **99**, 196805 (2007).

67. K. Sengupta, D. Sen, and S. Mondal, *Phys. Rev. Lett.* **100**, 077204 (2008).

68. J. T. Barreiro, M. Müller, P. Schindler, D. Nigg, T. Monz, M. Chwalla, M. Hennrich, C. F. Roos, P. Zoller, and R. Blatt, *Nature* **470**, 486 (2011).

69. M. Müller, K. Hammerer, Y. L. Zhou, C. F. Roos, and P. Zoller, *New J. Phys.* **13**, 085007 (2011).

70. J. Casanova, A. Mezzacapo, L. Lamata, and E. Solano, *Phys. Rev. Lett.* **108**, 190502 (2012).

71. K. Takahasi, *Phys. Rev. E* **87**, 062117 (2013).

72. H. J. Lipkin, N. Meshkov, and A. J. Glick, *Nucl. Phys.* **62**, 188 (1965).

73. S. Ducci, P. L. Ramazza, W. González-Viñas, and F. T. Arecchi, *Phys. Rev. Lett.* **83**, 5210 (1999).

74. S. Casado, W. González-Viñas, H. Mancini, and S. Boccaletti, *Phys. Rev. E* **63**, 057301 (2001).

75. S. Casado, W. González-Viñas, and H. Mancini, *Phys. Rev.* **74**, 047101 (2006).

76. S. Casado, W. González-Viñas, S. Boccaletti, P. L. Ramazza, and H. Mancini, *Eur. J. Phys. Spec. Top.* **146**, 87 (2007).

77. M. A. Miranda, J. Burguete, W. González-Viñas, and H. Mancini, *Int. J. Bifurcation and Chaos* **22**, 1250165 (2012).

78. M. A. Miranda, J. Burguete, H. Mancini, and W. González-Viñas, *Phys. Rev. E* **87**, 032902 (2013).

79. P. Ashcroft and T. Galla, arXiv:1308.6101 (2013).

80. J. R. Anglin and W. H. Zurek, *Phys. Rev. Lett.* **83**, 1707 (1999).

81. R. Monaco, J. Mygind, and R. J. Rivers, *Phys. Rev. Lett.* **89**, 080603 (2002).

82. R. Monaco, J. Mygind, and R. J. Rivers, *Phys. Rev. B* **67**, 104506 (2003).

83. R. Monaco, J. Mygind, M. Aaroe, R. J. Rivers, and V. P. Koshelets, *Phys. Rev. Lett.* **96**, 180604 (2006).
84. R. Monaco, M. Aaroe, J. Mygind, R. J. Rivers, and V. P. Koshelets, *Phys. Rev. B* **77**, 054509 (2008).
85. E. Kavoussanaki, R. Monaco, and R. J. Rivers, *Phys. Rev. Lett.* **85**, 3452 (2000).
86. R. Monaco, J. Mygind, R. J. Rivers, and V. P. Koshelets, *Phys. Rev. B* **80**, 180501(R) (2009).
87. W. H. Zurek, *J. Phys.: Condens. Matter* **25**, 404209 (2013); doi:10.1088/0953-8984/25/40/404209.
88. F. Liu and G. F. Mazenko, *Phys. Rev. B* **46**, 5963 (1992).
89. S. C. Chae, N. Lee, Y. Horibe, M. Tanimura, S. Mori, B. Gao, S. Carr, and S.-W. Cheong, *Phys. Rev. Lett.* **108**, 167603 (2012).
90. S. M. Griffin, M. Lilienblum, K. Delaney, Y. Kumagai, M. Fiebig, and N. A. Spaldin, *Phys. Rev. X* **2**, 041022 (2012).
91. M. Campostrini, M. Hasenbusch, A. Pelissetto, and E. Vicari, *Phys. Rev. B* **74**, 144506 (2006); A. Pelissetto and E. Vicari, *Phys. Rep.* **368**, 549 (2002).
92. A. del Campo, T. W. B. Kibble, and W. H. Zurek, *J. Phys.: Condens. Matter* **25**, 404210 (2013).
93. W. H. Zurek, *Phys. Rev. Lett.* **102**, 105702 (2009).
94. W. H. Zurek and U. Dorner, *Phil. Trans. R. Soc. A* **366**, 2953 (2008).
95. B. Damski and W. H. Zurek, *New J. Phys.* **11**, 063014 (2009).
96. J. Dziarmaga and M. M. Rams, *New J. Phys.* **12**, 055007 (2010).
97. J. Dziarmaga and M. M. Rams, *New J. Phys.* **12**, 0103002 (2010).
98. A. del Campo, A, G. De Chiara, G. Morigi. M. B. Plenio, and A. Retzker, *Phys. Rev. Lett.* **105**, 075701 (2010).
99. A. del Campo, A. Retzker, and M. B. Plenio, *New J. Phys.* **13**, 083022 (2011).
100. E. Witkowska, P. Deuar, M. Gajda, and K. Rzażewski, *Phys. Rev. Lett.* **106**, 135301 (2011).
101. M. Collura and D. Karevski, *Phys. Rev. Lett.*, **104**, 200601 (2010).
102. H. Haeffner, C. F. Roos, and R. Blatt, *Phys. Rep.* **469**, 155 (2008).
103. C. Schneider, D. Porras, and T. Schaetz, *Rep. Prog. Phys.* **75**, 024401 (2012).
104. R. Blatt and C. F. Roos, *Nature Phys.* **8**, 277 (2012).
105. G. Birkl, S. Kassner, and H. Walther, *Nature* **357**, 310 (1992).
106. I. Waki, S. Kassner, G. Birkl, and H. Walther, *Phys. Rev. Lett.* **68**, 2007 (1992).
107. D. H. E. Dubin and T. M. O'Neil, *Rev. Mod. Phys.* **71**, 87 (1999).
108. J. P. Schiffer, *Phys. Rev. Lett.* **70**, 818 (1993).
109. G. Piacente, I. V. Schweigert, J. J. Betouras, and F. M. Peeters, *Phys. Rev. B* **69**, 045324 (2004).
110. S. Fishman, G. De Chiara, T. Calarco, and G. Morigi, *Phys. Rev. B* **77**, 064111 (2008).
111. H. L. Partner, R. Nigmatullin, T. Burgermeister, K. Pyka, J. Keller, A. Retzker, M. B. Plenio, and T. E. Mehlstäubler, arXiv:1305.6773; to be published in *New. J. Phys.* (2013); doi:10.1088/1367-2630/15/10/103013.

112. H. Saito, Y. Kawaguchi, and M. Ueda, *Phys. Rev. A* **76**, 043613 (2007).
113. J. Dziarmaga, J. Meisner, and W. H. Zurek, *Phys. Rev. Lett.* **101**, 115701 (2008).
114. R. Nigmatullin, A. del Campo, G. De Chiara, G. Morigi, M. B. Plenio, and A. Retzker, arXiv:1112.1305 (2011).
115. E. Shimshoni, G. Morigi, and S. Fishman, *Phys. Rev. Lett.* **106**, 010401 (2011).
116. E. Shimshoni, G. Morigi, and S. Fishman, *Phys. Rev. A* **83**, 032308 (2011).
117. C. Cormick and G. Morigi, *Phys. Rev. Lett.* **109**, 053003 (2012).
118. J. D. Baltrusch, C. Cormick, and G. Morigi, *Phys. Rev. A* **86**, 032104 (2012).
119. J. D. Baltrusch, C. Cormick, and G. Morigi, arXiv:1301.3646 (2013).
120. J. Dziarmaga, W. H. Zurek, and M. Zwolak, *Nature Phys.* **8**, 49 (2012).
121. J. D. Baltrusch, C. Cormick, G. De Chiara, T. Calarco, and G. Morigi, *Phys. Rev. A* **84**, 063821 (2011).
122. M. M. Rams, M. Zwolak, and B. Damski, *Sci. Rep.* **2**, 655 (2012) .
123. H. Landa, S. Marcovitch, A. Retzker, M. B. Plenio, and B. Reznik, *Phys. Rev. Lett.* **104**, 043004 (2010).
124. L. Cincio, J. Dziarmaga, M. M. Rams, and W. H. Zurek, *Phys. Rev. A* **75**, 052321 (2007).
125. H. Landa, A. Retzker, T. Schaetz, and B. Reznik, arXiv:1308.2943 (2013).
126. B. Damski, H. T. Quan, and W. H. Zurek, *Phys. Rev. A* **83**, 062104 (2011).
127. B. Damski and W. H. Zurek, *Phys. Rev. Lett.* **104**, 160404 (2010).
128. T. Karpiuk, P. Deuar, P. Bienias, E. Witkowska, K. Pawlowski, M. Gajda, K. Rzazewski, and M. Brewczyk, *Phys. Rev. Lett.* **109**, 205302 (2012).
129. D. S. Petrov, G. V. Shlyapnikov, and J. T. M. Walraven, *Phys. Rev. Lett.* **85**, 3745 (2000).
130. A. Imambekov *et al.*, *Phys. Rev. A* **80**, 033604 (2009).
131. S. Manz, R. Bücker, T. Betz, Ch. Koller, S. Hofferberth, I. E. Mazets, A. Imambekov, E. Demler, A. Perrin, J. Schmiedmayer, and T. Schumm, *Phys. Rev. A* **81**, 031610(R) (2010).
132. J. G. Muga, Xi Chen, A. Ruschhaupt, and D. Guery-Odelin, *J. Phys. B* **42**, 241001 (2009).
133. A. del Campo, *Phys. Rev. A* **84**, 031606(R) (2011).
134. J.-F. Schaff, X.-L. Song, P. Vignolo, and G. Labeyrie, *Phys. Rev. A* **82**, 033430 (2010).
135. J.-F. Schaff, X.-L. Song, P. Capuzzi, P. Vignolo, and G. Labeyrie, *EPL* **93**, 23001 (2011).
136. T. Donner, S. Ritter, T. Bourdel, A. Öttl, M. Köhll, and T. Esslinger, *Science* **315**, 1556 (2007).
137. P. C. Hohenberg and B. I. Halperin, *Rev. Mod. Phys.* **49**, 435 (1977).
138. D. R. Scherer, C. N. Weiler, T. W. Neely, and B. P. Anderson, *Phys. Rev. Lett.* **98**, 110402 (2007).
139. M. P. A. Fisher, P. B. Weichman, G. Grinstein, and D. S. Fisher, *Phys. Rev. B* **40**, 546 (1989).
140. D. Jaksch, C. Bruder, J. I. Cirac, C. W. Gardiner, and P. Zoller, *Phys. Rev. Lett.* **81**, 3108 (1998).

141. M. Greiner, O. Mandel, T. Esslinger, T. W. Hänsch, and I. Bloch, *Nature (London)* **415**, 39 (2002).
142. W. S. Bakr, A. Peng, M. E. Tai, R. Ma, J. Simon, J. I. Gillen, S. Fölling, L. Pollet, and M. Greiner, *Science* **229**, 547 (2010).
143. R. Schützhold, M. Uhlmann, Y. Xu, and U. R. Fischer, *Phys. Rev. Lett.* **97**, 200601 (2006).
144. F. M. Cucchietti, B. Damski, J. Dziarmaga, and W. H. Zurek, *Phys. Rev. A* **75**, 023603 (2007).
145. J. Dziarmaga, J. Meisner, and W. H. Zurek, *Phys. Rev. Lett.* **101**, 115701 (2008).
146. J. Dziarmaga, M. Tylutki, and W. H. Zurek, *Phys. Rev. B* **86**, 144521 (2012).
147. M. Tylutki, J. Dziarmaga, and W. H. Zurek, *J. Phys.: Conf. Ser.* **414**, 012029 (2013).
148. D. Chen, M. White, C. Borries, and B. DeMarco, *Phys. Rev. Lett.* **106**, 235304 (2011).
149. D. A. R. Dalvit, J. Dziarmaga, and W. H. Zurek, *Phys. Rev. A* **62**, 013607 (2000).
150. P. Laguna and W. H. Zurek, *Phys. Rev. D* **58**, 8752 (1998).
151. A. Yates and W. H. Zurek, *Phys. Rev. Lett.* **80**, 5477 (1998).
152. M. Hindmarsh and A. Rajantie, *Phys. Rev. Lett.* **85**, 4660 (2000).
153. G. J. Stephens, L. M. A. Bettencourt, and W. H. Zurek, *Phys. Rev. Lett.* **88**, 137004 (2002).
154. T. W. B. Kibble and A. Rajantie, *Phys. Rev. B* **68**, 174512 (2003).
155. T. Caneva, R. Fazio, and G. E. Santoro, *Phys. Rev. B* **76**, 144427 (2007).
156. F. Pellegrini, S. Montangero, G. E. Santoro, and R. Fazio, *Phys. Rev. B* **77**, 140404(R) (2008).
157. A. Bermudez, D. Patanè, L. Amico, and M. A. Martin-Delgado, *Phys. Rev. Lett.* **102**, 135702 (2009).
158. S. Deng, G. Ortiz, and L. Viola, *Phys. Rev. B* **83**, 094304 (2011).
159. J. Sabbatini, W. H. Zurek, and M. J. Davis, *Phys. Rev. Lett.* **107**, 230402 (2011).
160. J. Sabbatini, W. H. Zurek, and M. J. Davis, *New J. Phys.* **14**, 095030 (2012).
161. A. Jelic and L. F. Cugliandolo, *J. Stat. Mech.* P02032 (2011).
162. G. Vacanti, S. Pugnetti, N. Didier, M. Paternostro, G. M. Palma, R. Fazio, and V. Vedral, *Phys. Rev. Lett.* **108**, 093603 (2012).
163. H. Saito, Y. Kawaguchi, and M. Ueda, *J. Phys.: Condens. Matter* **25**, 404212 (2013).
164. M. A. Miranda, D. Laroze, and W. González-Viñas, *J. Phys.: Condens. Matter* **25**, 404208 (2013).
165. S.-W. Su, S.-C. Gou, A. Bradley, O. Fialko, and J. Brand, *Phys. Rev. Lett.* **110**, 215302 (2013).
166. T. Świsłocki, E. Witkowska, J. Dziarmaga, and M. Matuszewski, *Phys. Rev. Lett.* **110**, 045303 (2013).
167. A. Lamacraft, *Phys. Rev. Lett.* **98**, 160404 (2007).
168. M. Uhlmann, R. Schützhold, and U. R. Fischer, *Phys. Rev. Lett.* **99**, 120407 (2007).

169. A. Chandran, A. Erez, S. S. Gubser, and S. L. Sondhi, *Phys. Rev. B* **86**, 064304 (2012).
170. A. Chandran, F. J. Burnell, V. Khemani, and S. L. Sondhi, *J. Phys.: Condens. Matter* **25**, 404214 (2013).
171. M. Oberthaler, private communication.
172. E. Timmermans, *Phys. Rev. Lett.* **81**, 5718 (1998).
173. R. Islam, E. E. Edwards, K. Kim, S. Korenblit, C. Noh, H. Carmichael, G.-D.Lin, L.-M. Duan, C.-C. Joseph Wang, J. K. Freericks, and C. Monroe, *Nat. Commun.* **2**, 377 (2011).
174. C. Schneider, D. Porras, and T. Schaetz, *Rep. Prog. Phys.* **75**, 024401 (2012).
175. R. Yusupov, T. Mertelj, V. V. Kabanov, S. Brazovskii, P. Kusar, J.-H. Chu, I. R. Fisher, and D. Mihailovic, *Nature Phys.* **6**, 681 (2010).
176. D. Mihailovic, T. Mertelj, V. V. Kabanov, and S. Brazovskii, arXiv: 1304.6968 (2013).
177. H. Ikegami, Y. Tsutsumi, and K. Kono *Science* **341**, 59 (2013).
178. S. Coleman, *Aspects of Symmetry: Selected Erice Lectures* (Cambridge University Press, Cambridge, 1988).
179. H. Arodz, J. Dziarmaga, and W. H. Zurek, *Patterns of Symmetry Breaking* (Kluwer Academic Publishers, London, 2003).
180. N. D. Mermin, *Rev. Mod. Phys.* **51**, 591 (1979).
181. M. Ueda, *Fundamentals and New Frontiers of Bose–Einstein Condensation* (World Scientific, Singapore, 2010).
182. Y. Kawaguchi and M. Ueda, *Phys. Rep.* **520**, 253 (2012).
183. D. M. Stamper-Kurn and M. Ueda, *Rev. Mod. Phys.* **85**, 1191 (2013)
184. T. W. B. Kibble, G. Lazarides, and Q. Shafi, *Phys. Rev. D* **26**, 435 (1982).
185. M. Nakahara, *Geometry, Topology and Physics* (Taylor & Francis, New York, 2003).
186. G. E. Volovik, *JETP Lett.* **28**, 59 (1978).
187. V. P. Mineyev and G. E. Volovik, *Phys. Rev. B* **18**, 3197 (1978).

THE QUEST FOR THE HIGGS BOSON AT THE LHC

TEJINDER S. VIRDEE

Blackett Laboratory, Imperial College, London, UK

In July 2012 the ATLAS and CMS experiments announced the discovery of a Higgs boson, confirming the conjecture put forward by Tom Kibble and others in the 1960s. This article will attempt to outline some of the challenges faced during the construction of the Large Hadron Collider and its experiments, their operation and performance, and selected physics results. In particular, results relating to the new heavy boson will be discussed as well as its properties and the future prospects for the LHC programme.

1. Introduction

It is a great honour for me to speak at the celebration of Tom Kibble's 80th birthday. Tom has made remarkable contributions to the elucidation of the Standard Model (SM) of particle physics. The SM that has emerged over the last five decades is based upon principles of great beauty and simplicity. It comprises the building blocks of visible matter, the fundamental fermions: quarks and leptons, and the fundamental bosons that mediate three of the four fundamental interactions; photons for electromagnetism, the W and Z bosons for the weak interaction and gluons for the strong interaction (Fig. 1). The photon is massless whilst the W and Z bosons acquire mass through a spontaneous symmetry-breaking mechanism proposed by three groups of physicists (Englert and Brout; Higgs; and Guralnik, Hagen, and Kibble).[1-5] This is achieved through the introduction of a complex scalar field leading to an additional massive scalar boson, labeled the SM Higgs boson. The fundamental fermions acquire mass through a Yukawa interaction of the Higgs boson. Only the gravitational interaction remains outside the SM.

This paper shall give a brief historical perspective on the quest at the Large Hadron Collider (LHC) for the SM Higgs boson, spanning the last two decades. In the early 1990s the search for the SM Higgs boson played a pivotal role in the design of the ATLAS and CMS experiments.

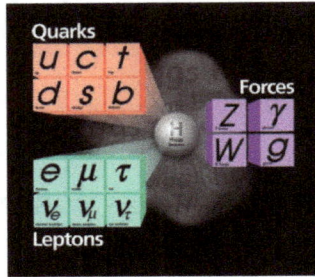

Fig. 1.   The particle content of the SM.

In July 2012 the ATLAS and CMS collaborations discovered a Higgs boson.[10,11] In the list of references in these discovery papers the first were the five seminal papers published some fifty years ago that introduced spontaneous symmetry breaking.[1-5] The sixth paper in this list, also a seminal one, is authored by Kibble in 1967.[6] It is in this paper that Kibble generalized his earlier work with Guralnik and Hagen and brought the mechanism of spontaneous symmetry breaking closer to the application to the real world, one in which the photon remains massless and the W and Z particles become massive[12] with masses of 80.4 GeV and 91.2 GeV, respectively. The three papers[7-9] that follow in the list of references, again seminal, elucidate the theory of electroweak unification. Kibble's paper of 1967 influenced the work of Steven Weinberg and Abdus Salam on electroweak unification.

2. The Standard Model

Fundamental matter particles carry spin 1/2 whilst force mediator particles carry spin 1. The SM is a gauge quantum field theory containing the internal symmetries of the unitary product group SU(3) × SU(2) × U(1) and its Lagrangian can be written as:

$$L = -\frac{1}{4}F_{\mu\nu}F^{\mu\nu} + i\bar{\psi}\slashed{D}\psi + \text{h.c.} + \psi_i y_{ij}\psi_j\phi + \text{h.c.} + |D\phi|^2 + V(\phi).$$

The first term in the Lagrangian accounts for the gauge fields, the second for the fermion fields, the third for the interactions between the gauge fields and the fermions whereas the last term contains the scalar field and the Higgs boson terms.

The SM is a very successful description of our visible universe and has been verified in many experiments to a very high precision. It has an enormous range of applicability and validity. So far no significant deviations have been observed experimentally. However, as we embarked upon the LHC adventure in the 1980s there were many profound open questions in particle physics. In popular terms these are: what is the origin of mass i.e. how do fundamental particles acquire mass, and indeed why do they have the masses that they have? Why is there more matter than anti-matter? What constitutes dark matter? What is the path towards unification? Do we live in a world with more space-time dimensions than the familiar four? The LHC[13] was conceived to address or shed light on these questions in particle physics.

A major goal of the LHC was the elucidation of the mechanism for electroweak symmetry breaking.

It is well known that the SM contains too many arbitrary parameters (e.g couplings and masses are put in by hand) whose values would presumably be predicted in a unified theory.

Quantum corrections make the mass of a fundamental scalar particle float up to the next highest physical mass scale that, in the absence of extensions to the SM, is as high as 10^{15} GeV. Hence finding the Higgs boson would immediately raise a more puzzling question: Why should it have a mass in the mass range explored by the LHC i.e. between 100 GeV and 1 TeV? It was widely believed that the answer to this question would lie in new physics beyond the SM (BSM). One appealing hypothesis, much discussed at the time, and still being investigated, predicts a new symmetry labeled *supersymmetry*. For each known SM particle there would be a partner with spin differing by half a unit; fermions would have boson partners and vice versa, thus doubling the number of fundamental particles. The contributions from the boson and fermion superpartners, and vice versa, would lead to the cancellations and allow a low mass for the Higgs boson. In the simplest forms of supersymmetry five Higgs bosons are predicted to exist with one resembling the SM Higgs boson with a mass below ~140 GeV. The lightest of this new species of super-particles could be the candidate for *dark matter* in the universe that is around five times more abundant than ordinary matter.

Finally it was clear that a search had to be made for new physics at the TeV energy scale as the SM is logically incomplete; it does not incorporate gravity. Superstring theory is an attempt towards a unified theory with dramatic predictions of extra space dimensions and supersymmetry.

When ATLAS and CMS started recording high energy proton-proton collision data in 2010, within a month or so plots such as Fig. 2 were

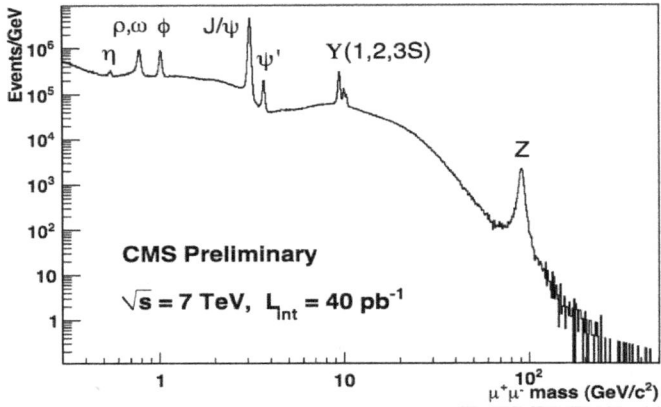

Fig. 2. The distribution of di-muon effective masses showing the various resonant states.

produced which effectively illustrated 50 years of particles physics; the peaks from left to right represent the η, ρ, ω and ϕ articles discovered in the 1960s; the J/ψ – a charm quark-antiquark pair discovered in 1974, the Y – a bottom quark-antiquark pair discovered in 1979, and the Z boson discovered in 1983. The top quark, not in Fig. 2, was found in the mid 1990s, almost completing the particle content of the standard model (SM) of particle physics illustrated in Fig. 1. The only missing element was the Higgs boson.

In 1975, physicists had already started to turn their attention to how a putative Higgs boson would manifest itself in experiments, were they to be performed.[14] Today, the concluding paragraph in this paper makes for interesting reading:

"We should perhaps finish with an apology and a caution. We apologize to experimentalists for having no idea what is the mass of the Higgs boson, unlike the case with charm and for not being sure of its couplings to other particles, except that they are probably all very small. For these reasons we do not want to encourage big experimental searches for the Higgs boson, but we feel that people performing experiments vulnerable to the Higgs boson should know how they turn up."

Some 40 years later and in a multi-billion Swiss francs project ATLAS and CMS experiments have discovered a new boson that resembles very much the SM Higgs boson.

This year (2013) is also the 30th anniversary of the discovery of the W and Z bosons by the UA1 and UA2 experiments at the proton-antiproton collider at CERN. The discovery of the W and Z bosons focused efforts, and set the stage, for the search for the Higgs boson. In 1984, the following year, a workshop was held in Lausanne where first ideas were discussed about a possible proton-proton collider and associated experiments for this search. The aim was to reuse the LEP tunnel after the end of the electron-positron programme. Amongst the leading protagonists were the scientists from UA1 and UA2 experiments. An exploratory machine was required to cover the wide range of mass, the diverse signatures and mechanisms thought to be effective for the production of the new particles at a centre-of-mass energy ten times higher than previously probed. A hadron (proton–proton) collider is such a machine as long as the proton energy is high enough and the instantaneous luminosity, L, is sufficiently large. L is measured in $\mathrm{cm}^{-2}\,\mathrm{s}^{-1}$, and is indicative of the number of proton-proton collisions taking place per second. The predicted rate of production of a given particle is given by $L \times \sigma$ where σ is the cross section of the production reaction, measured in units of cm^2. The most interesting and easily detectable final states at a hadron collider involve charged leptons and photons and have a low $\sigma \times BR$, where BR is the branching ratio into the decay mode of interest. The hadron colliders can provide these conditions though at the expense of "clean" experimental conditions.

The LHC and its experiments were designed to find new particles, new forces and new symmetries amongst which could be the Higgs boson(s), supersymmetric particles, Z' bosons, or evidence of extra space dimensions. An experiment that could cover the detection of all these "known" but yet undiscovered particles would also allow discovery of whatever else Nature has in store at the LHC energies.

3. The Standard Model Higgs Boson and the LHC

The mass of the Higgs boson is not predicted by theory but depending on its mass all its other properties are precisely predicted. The SM Higgs boson is short-lived (10^{-23} s) and hence the experiments would only detect its decay products.

From general considerations $m_H < 1$ TeV whilst precision electroweak constraints imply that $m_H < 152$ GeV at 95% confidence level (CL).[15] The lower limit on the mass of the Higgs boson from the LEP experiments was 114.4 GeV.[16]

The production cross sections and the branching ratios into the various decay modes of the SM Higgs boson as a function of mass are illustrated in Figs. 3(a) and 3(b), respectively.[17] The dominant Higgs-boson

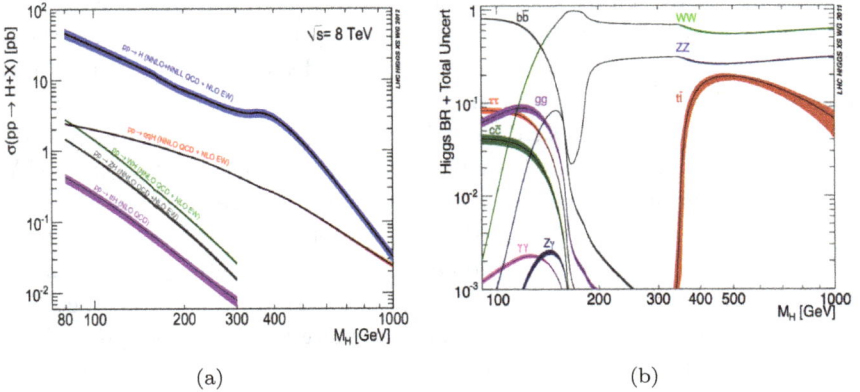

(a)   (b)

Fig. 3.   (a) The SM Higgs production cross-section at $\sqrt{s} = 8$ TeV. (b) The SM Higgs branching ratios as a function of the Higgs-boson mass.

production mechanism, for masses up to ≈ 700 GeV, is gluon-gluon fusion. The W–W or Z–Z fusion mechanism, known as vector boson fusion (VBF), becomes important for the production of higher-mass Higgs bosons. Here, the quarks that emit the W/Z bosons end up in the final states with transverse momenta of the order of W and Z masses. The detection of the resulting high-energy jets in the forward regions, $2.0 < |\eta| < 5.0$, can be used to tag the reaction, improving the signal-to-noise ratio and extending the mass range over which the Higgs can be discovered. The pseudorapidity $\eta = \ln[\tan(\theta/2)]$ where θ is the polar angle measured from the positive z axis (along the anticlockwise beam direction). These jets are highly boosted and their transverse size is similar to that of a high-energy hadron shower. The tagging of forward jets has turned out to be very important in the measurements for the newly found boson as well.

Once produced the Higgs boson disintegrates immediately in one of several ways (decay modes) into known SM particles, depending on its mass. Some of the decay modes are listed in Table 1. A search had to be envisaged not only over a large range of masses but also many possible decay modes: into pairs of photons, Z bosons, W bosons, τ leptons, and b quarks.

The search for the SM Higgs boson provided a stringent benchmark for evaluating the physics performance of various experiment designs under consideration some twenty years ago and heavily influenced the conceptual design of the general-purpose experiments, ATLAS and CMS. In the mass interval $110 < m_H < 150$ GeV, early detailed studies indicated that the two-photon decay would be the main channel likely to give a significant signal.[18] Detailed studies of another mode, $H \rightarrow ZZ^{(*)} \rightarrow \ell\ell\,\ell\ell$, labeled the "golden" mode, suggested that also it could be used to cleanly detect the Higgs boson over a wide range of masses starting around $m_H = 125$ GeV.[19] One or both of the Z's may be virtual in the range $120 < m_H < 180$ GeV and in the range $2m_Z < m_H < 600$ GeV both Z bosons are real. The Higgs boson should be detectable via its decay into two W bosons in the mass range $130 < m_H < 180$ GeV. The leptonic decays of the W and Z are used when possible. In the region $700 < m_H < 1000$ GeV the cross-section decreases so that higher branching ratio modes of the W or Z, involving jets or transverse missing momentum (or commonly labeled missing transverse energy, E_T^{miss}) have to be employed.

Table 1. The promising decay modes of the SM Higgs boson.

Region 1: Low mass region (LEP limit 114.5 GeV $< m_H < 2m_Z$; $m_Z = 91.2$ GeV)
$m_H < 150$ GeV: $H \rightarrow \gamma\gamma$, $Z\gamma$
120 GeV/c^2 $< m_H < 2m_Z$: $H \rightarrow ZZ^* \rightarrow \ell\ell\,\ell\ell$
120 GeV/c^2 $< m_H < 2m_Z$: $H \rightarrow WW^* \rightarrow \ell\nu\,\ell\nu$
120 GeV/c^2 $< m_H < 2m_Z$: $qq \rightarrow qqH$ with $H \rightarrow \tau\tau$, $H \rightarrow WW$ etc.
$m_H < 130$ GeV: $pp \rightarrow WH \rightarrow \ell\nu\,b\bar{b}$ or $t\bar{t}H \rightarrow \ell\nu X\,b\bar{b}$
Region 2: High mass region ($2m_Z < m_H < 700$ GeV)
$H \rightarrow ZZ \rightarrow \ell\ell\,\ell\ell$
Region 3: Very high mass region ($700 < m_H < 1$ TeV)
$H \rightarrow ZZ \rightarrow \ell\ell\,\nu\nu$, $H \rightarrow ZZ^* \rightarrow \ell\ell$ jet–jet
$H \rightarrow WW \rightarrow \ell\nu$ jet–jet

4. The Large Hadron Collider Project

Hadron colliders are "broad-band" exploratory machines in the sense that a wide range of centre-of-mass energies can be explored simultaneously as the fraction of the proton energy carried by the constituents varies from interaction to interaction.

Not only is the putative SM Higgs boson rarely produced in the proton collisions, it also rarely decays into particles that are the best identifiable signatures of its production at the LHC: photons, electrons, and muons. The rarity is illustrated by the fact that Higgs boson production and decay to one such distinguishable signature ($H \rightarrow ZZ \rightarrow 4l$) happens roughly only once in 10 trillion proton-proton collisions. This means that a large number of proton-proton collisions per second have to be studied, the operating number at the end of 2012 being around 600 million per second, corresponding to an instantaneous luminosity of 7.7×10^{33} cm^{-2} s^{-1}. Hence the ATLAS and CMS detectors operate in the harsh environment created by this huge rate of proton-proton collisions.

4.1. Timeline of the LHC project

The long duration of the LHC Project so far is illustrated in Table 2. In the late 1980s and early 1990s several workshops and conferences took place where the formidable experimental challenges[20] started to appear manageable, provided that enough R&D work, especially on detectors, could be

Table 2.   The Timeline of the LHC project.

1984	Workshop on a Large Hadron Collider in the LEP tunnel, Lausanne.
1987	Workshop on Physics at Future Accelerators, La Thuile, Italy. The Rubbia "Long-Range Planning Committee" recommends the Large Hadron Collider as the right choice for CERN's future.
1990	European Committee for Future Accelerators (ECFA) LHC Workshop, Aachen (discussion of physics, technologies and designs for LHC experiments)
1992	General Meeting on LHC Physics and Detectors, Evian les Bains (4 general-purpose experiment designs presented along with their physics performance)
1993	Three Letters of Intent submitted to the CERN peer review committee LHCC. ATLAS and CMS selected to proceed to a detailed technical proposal.
1994	The LHC accelerator approved for construction
1996	ATLAS and CMS Technical Proposals approved.
1997	Formal approval for ATLAS and CMS to move to construction (materials cost ceiling of 475 MCHF)
1997	Construction commences (after approval of detailed engineering design of subdetectors: magnets, inner tracker, calorimeters, muon system, trigger and data acquisition)
2000	Assembly of experiments commences, LEP accelerator is closed down to make way for the LHC.
2008	LHC experiments ready for pp collisions. LHC starts operation. An incident stops LHC operation.
2009	LHC restarts operation, pp collisions recorded by LHC detectors
2010	LHC collides protons at high energy (centre of mass energy of 7 TeV)
2012	LHC operates at 8 TeV: discovery of a Higgs boson.

carried out. In the early 1990s CERN setup a committee, the Detector R&D Committee (DRDC), to oversee research and development of technologies required to accomplish the physics goals of the LHC. It reviewed and steered R&D collaborations and greatly stimulated innovative developments in detector technology.

At the Aachen meeting, in 1990, discussions focused on the physics potential, the detector technologies and magnetic field configurations to deploy in possible experiments. The natural width of the SM Higgs boson in the low mass region is very small; it is < 10 MeV. The natural width is given by $\Gamma = \hbar/\tau$, where τ is the lifetime and $\hbar = h/2\pi$, h being the Planck's constant. Hence the width of any observed peak would be entirely dominated by instrumental mass-resolution. Considerable emphasis was therefore put on the value of the magnetic field strength, on the precision charged particle tracking systems and on high-resolution electromagnetic calorimeters. The high-mass Higgs boson region and signatures from supersymmetry drove the need for good resolution for the measurement of the energies of jets and missing transverse energy, as well as for almost full geometric coverage by the calorimeters.

At the Evian meeting in 1992 four experiment designs were presented: two deploying toroids (one superconducting) and two deploying superconducting high-field solenoids. Subsequently, lively discussions took place in the community on the possibility of joining forces. In June 1993 CERN's scientific peer review committee, the LHC Committee (LHCC), recommended that ATLAS and CMS experiments proceed further to the phase of technical proposals. The CMS design[21] was based on a single large-bore, long, high-field solenoid for analyzing muons, together with powerful microstrip-based inner tracking and an electromagnetic calorimeter of scintillating crystals. ATLAS,[22] in a complementary design, centred on a very large superconducting air-core toroid for the measurement of muons, supplemented by a superconducting 2 Tesla solenoid to provide the magnetic field for inner tracking and by a liquid-argon/lead electromagnetic calorimeter with a novel "accordion" geometry.

A saying prevalent in the late 1980s and early 1990s captured the challenge: "We think we know how to build a high energy, high luminosity hadron collider – but we don't have the technology to build a detector for it". Many technical, financial, industrial and human challenges lay ahead which were all overcome to yield experiments of unprecedented complexity and power. A flavour can be attained from articles in Ref. 23.

In the 1990s the two collaborations grew most rapidly in terms of people and institutes. Finding new collaborators was high on the "to do" list of the leaders of the experiments.

The formal approval for construction was given in July 1997 by the then director-general, Chris Llewellyn Smith, imposing a material cost ceiling of 475 MCHF. An intense ten-year period of construction ensued. In 2008 the LHC experiments were ready for pp collisions.

5. The LHC Accelerator

In the LHC[13] protons are accelerated by powerful electric fields generated in superconducting r.f. cavities and are guided around their circular orbits by powerful superconducting dipole magnets. The dipole magnets operate at 8.3 Tesla and 1.9 K in superfluid helium. The protons travel in a tube that is under a better vacuum, and at a lower temperature, than that found in inter-planetary space. The counter-rotating LHC beams are organized in 2808 bunches comprising $>10^{11}$ protons per bunch separated by 25 ns leading to a bunch crossing rate of ~ 40 MHz (up to now the LHC accelerator has operated at 50 ns bunch spacing with 1380 bunches). The main challenges for the accelerator were to build ~ 1200 15m-long superconducting dipoles able to reach this magnetic field, the large distributed cryogenic plant to cool the magnets and other accelerator structures, and the control of the beams.

Liquid helium displays two phenomena that are critical to the design of LHC: superconductivity and superfluidity. Some 100 years ago, Kamerling Onnes first liquefied Helium in his laboratory in Leiden at the rate of 60 ml per hour. The LHC today liquefies 32000 liters of liquid helium per hour in eight big cryogenic plants making this the largest refrigerator in the world.

The LHC magnets are cooled with pressurized superfluid helium that has unique engineering properties:

− a low bulk viscosity that allows the superfluid helium to permeate the smallest cracks. It is used to advantage in the magnet design by making the coil insulation slightly porous enabling the fluid to be in intimate contact with the superconductor,
− a large specific heat that is 100,000 times that of the superconductor per unit mass and 2000 times larger per unit volume and
− a thermal conductivity that peaks at 1.9 K and is ~ 1000 times that of good quality copper.

The stored energy in each of the beams at nominal intensity and energy is 350 MJ, equivalent to more than 80 kg of TNT. Hence if the beam is lost in an uncontrolled way it can do considerable damage to the machine components that would result in months of down-time.

6. The ATLAS and CMS Experiments

The typical form of a collider detector is a "cylindrical onion" comprising four principal layers instrumented with technologically advanced detectors covering the full (4π) solid angle. A photograph of the mid-plane transverse cut is very illustrative (Fig. 4). The detectors in each layer are designed to perform a specific task and all together these layers allow the identification and precise measurement of the energies and directions of all the particles produced in proton-proton collisions.

Fig. 4. Transverse section of the barrel part of CMS illustrating the successive layers of detection starting from the centre where the collisions occur: the inner tracker, the crystal calorimeter, the hadron calorimeter, the superconducting coil, and the iron yoke instrumented with the four muon stations. The last muon station is at a radius of 7.4 m.

A particle emerging from the collision and traveling outward will first encounter the inner tracking system, immersed in a uniform magnetic field, comprising an array of pixels and microstrip detectors. These precisely measure the trajectory of the spiraling charged particles and the curvature of their paths, revealing their momentum. The stronger the magnetic field, the higher the curvature of the paths, and the more precise is the measurement of the particle's momentum. The energies of particles are measured in the next two layers of the detector, the electromagnetic and hadronic calorimeters. The energies of electrons and photons are measured by the electromagnetic calorimeter, and of jets by the electromagnetic and hadronic calorimeters. The only known particles that penetrate beyond the hadron calorimeter are muons and neutrinos. Muons, being charged particles, are tracked in dedicated muon chambers. Their momenta are also measured from the curvature of their paths in a magnetic field. Neutrinos escape detection and their presence gives rise to E_T^{miss}.

In order to discover the phenomena mentioned above protons collide head-on in what is termed a "hard interaction" as opposed to a glancing collision where less energy is involved in the physics of the interaction. In a hard interaction it is the constituents of the protons (quarks and gluons) that collide head-on. Any new particles produced as a result of these "hard" interactions will manifest themselves through the known and well-studied particles of the SM mentioned above. The photons, electrons and muons can emerge into the detectors directly from the hard interaction, whereas quarks and gluons, never visible as free particles, appear in the detectors as collimated bunches of stable or quasi-stable particles labeled "jets" (Fig. 5).

To accomplish the physics goals of the LHC new detector technologies had to be invented and most of the existing ones had to be pushed to their limits. Below we focus on the CMS experiment.

The construction of CMS commenced in 1998 and was completed early August 2008. CMS is the first experiment of its kind to be mostly assembled in a large surface hall in 15 slices and then meticulously lowered element-by-element, through a 23m-diameter shaft, by a huge custom-built gantry crane, into the underground cavern some 100 metres underground. A spectacular operation (Fig. 6) was the lowering of the central and heaviest slice in February 2007. The installation and the commissioning of the experiment finished in August 2008. On the 10th of September 2008 first beams circulated in the Large Hadron Collider. Nine days later, during the powering test of the last octant, alarms reached the LHC accelerator's control room and safety systems were activated to protect the accelerator. It turned out

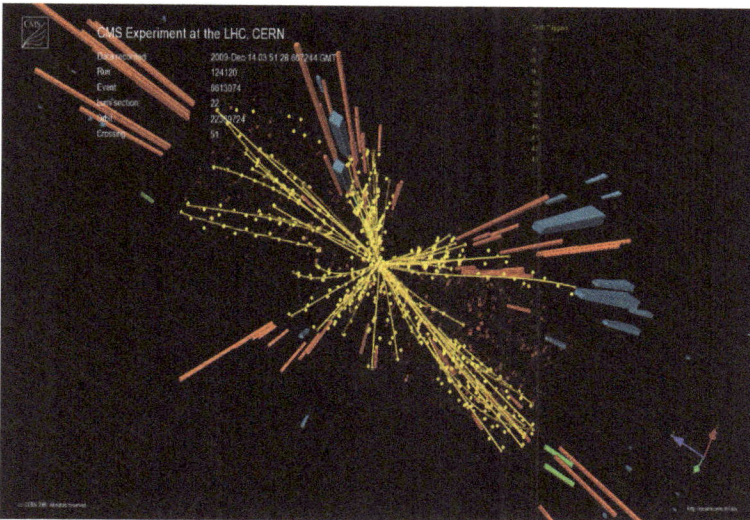

Fig. 5.   An event display of a proton-proton collision in CMS illustrating the inner tracking detector hits and the energy deposits in the electromagnetic calorimeter (red) and the hadron calorimeter (blue). The height of the red and blue towers is indicative of the energy deposited. The muon system is not displayed. The production of "jets" is evident from the collimated bunches of particles.

Fig. 6.   Lowering of the central element of CMS into the underground experiment cavern. It weighs ~2000 tons and took some ten hours to lower from the surface.

that one of the 50000 soldered joints had malfunctioned. This led to an electrical arc that pierced the vacuum enclosure of a superconducting dipole bending magnet leading a massive escape of helium, the pressure wave of which caused considerable collateral damage. The accelerator went offline and nine months of repairs ensued. The CMS experiment continued to run round-the-clock for a few months recording billions of traversals of muons from cosmic rays. These data demonstrated that the experiment was in a good shape to take collision data. After a few tweaks the CMS experiment was even better prepared for first collisions, which came on 22nd November 2009. The first collision data were rapidly distributed, analysed and physics results produced.

The CMS experiment started recording high-energy proton-proton collisions in March 2010 following a preliminary low-energy run in the autumn of 2009. In 2010, at $\sqrt{s} = 7$ TeV some 45 pb^{-1} of data were recorded, sufficient to demonstrate that not only the CMS experiment was working according to the ambitious design specifications of the conceivers, but also that it was producing known physics that was consistent with the predictions of the SM. An example of the performance can be seen from Fig. 2 where the width of the particles such as Y, dominated by instrumental resolutions, is measured to be 70 MeV consistent with the design resolution. More importantly CMS was now ready to delve into the unknown and look for new physics.

7. "Rediscovering" the Standard Model

In 2010 and 2011 CMS (and ATLAS) recorded an integrated luminosity of ~ 45 pb^{-1} and ~ 6 fb^{-1} respectively. Each fb^{-1} corresponds to the examination of ~ 80 trillion proton-proton collisions. These data allowed numerous precise measurements of SM processes including inclusive production of quarks (seen as hadronic jets), bottom-quarks, top-quarks and W and Z bosons, have been measured with high precision. The example in Fig. 7 compares the measurements of the cross section of various electroweak and electroweak+QCD processes with the predictions of the SM. The numerous and diverse measurements, in a previously unexplored energy region, confirm the predictions of the SM. It is essential to establish this agreement before any claims for new physics can be made, as SM processes constitute large backgrounds to new physics.

Using all the data so far collected extensive searches for new physics, beyond the standard model, have been performed. No new physics beyond the SM has yet been discovered. Limits have been set on quark substructure,

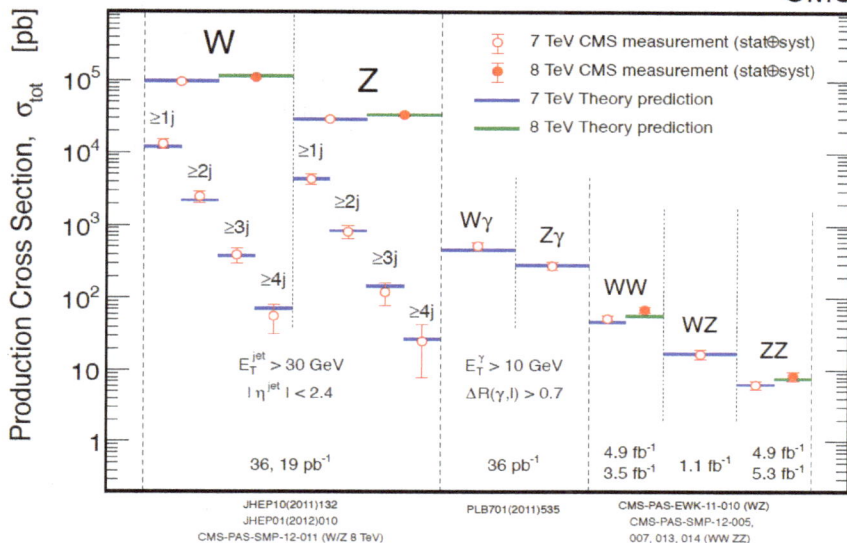

Fig. 7. A comparison of the measurements of electroweak and QCD processes with the predictions of the SM.

supersymmetric particles (e.g. disfavouring gluino masses below 1 TeV in simple models of supersymmetry), potential new bosons (e.g disfavouring new heavy W' and Z' bosons with masses below 2 TeV for couplings similar to the ones for the known W and Z bosons) and even signs of TeV-scale gravity (e.g. disfavouring black holes with masses below 4 TeV).

8. The Discovery and Properties of a Higgs Boson

Undoubtedly, the most striking result to emerge from the ATLAS[10] and CMS[11] experiments is the discovery of a new heavy boson with a mass of $\sim$125 GeV. The analysis was carried out in the context of the search for the standard model (SM) Higgs boson.

The predicted rate of production of the SM Higgs bosons, its decay modes and its natural width vary widely over the allowed mass range (100–1000 GeV). It couples to the different pairs of particles in a proportion that is precisely predicted by the SM i.e. for fermions (f) proportional to m_f^2 and for bosons (V) proportional m_V^4/v^2, where v is the vacuum expectation value of the scalar field $(v = 246$ GeV). Once produced the Higgs boson disintegrates immediately into known SM particles. A search had to be envisaged not only over a large range of masses but also many possible decay modes with differing branching ratios. For example, at $m_H = 125$ GeV the SM boson is predicted to decay into pairs of photons with BR $= 2.3 \times 10^{-3}$, into Z bosons and then four electrons or muons or two muons and two electrons with BR $= 1.25 \times 10^{-4}$, into a pair of W bosons and then into $ll\nu\nu$ with BR $\sim 1\%$, a pair of τ-leptons with BR $= 6.4\%$, and into a pair of b-quarks with BR $= 54\%$.

For a given Higgs boson mass hypothesis, the sensitivity of the search depends on:

– the mass of the Higgs boson
– the Higgs boson production cross section (Fig. 3(a)),
– the decay branching fraction into the selected final state (Fig. 3(b)),
– the signal selection efficiency,
– the observed Higgs boson mass resolution, and
– the level of backgrounds with the same or a similar final state

In the 2011 data-taking run the ATLAS and CMS experiments recorded data corresponding to an integrated luminosity of ~ 5 fb^{-1} at $\sqrt{s} = 7$ TeV. In December 2011, the very first "tantalizing hints" of a new particle from both the CMS and ATLAS experiments were shown at CERN. The general conclusion was that both experiments were seeing an excess of unusual events at roughly the same place in mass (in the mass range 120–130 GeV) in two different decay channels. That set the stage for data taking in 2012.

In January 2012 it was decided to slightly increase the energy of the protons from 3.5 to 4 TeV, giving a centre of mass energy of 8 TeV. By June 2012 the number of high-energy collisions examined had doubled and both CMS and ATLAS had greatly improved their analyses so it was decided to look at the area which had shown the excess of events but only after all the algorithms and selection procedures had been agreed, in case a bias was inadvertently introduced. These data led to the discovery of a Higgs boson, independently in both the ATLAS and CMS experiments in July 2012 (see Sec. 8.1).

In what follows we shall concentrate on the region of low mass (114 < m_H < 150 GeV) where the two channels particularly suited for unambiguous discovery are the decays to two photons and to two Z bosons, where one or both of the Z bosons could be virtual, subsequently decaying into four electrons, four muons or two electrons and two muons. These channels are particularly suitable as the observed mass resolution ($\sim 1\%$ of m_H) is the best and the backgrounds manageable or small.

By the end of 2012 (LHC Run 1) the total amount of data that had been examined corresponded to $\sim$5 fb^{-1} at $\sqrt{s} = 7$ TeV and $\sim$20 fb^{-1} at $\sqrt{s} = 8$ TeV, equating to the examination of some 2000 trillion proton-proton collisions. Using these data first measurement of the properties of the new boson also were made (see Sec. 8.2).

8.1. Results from the 2011 and partial 2012 datasets

In this section we discuss the analyses that led to the discovery of a new heavy boson around a mass of 125 GeV using the data accumulated up to June 2012.

8.1.1. The $H \to \gamma\gamma$ decay mode

In the $H \to \gamma\gamma$ analysis a search is made for a narrow peak in the diphoton invariant mass distribution in the mass range 110–150 GeV, on a large irreducible background from QCD production of two photons (via quark-antiquark annihilation and "box" diagrams). There is also a reducible background where one or more of the reconstructed photon candidates originate from misidentification of jet fragments, with the process of QCD Compton scattering dominating.

The event selection requires two photon candidates satisfying p_T and photon identification criteria. In CMS[10] a p_T threshold of $m_{\gamma\gamma}/3$ ($m_{\gamma\gamma}/4$) is applied to the photon leading (sub-leading) in p_T, where $m_{\gamma\gamma}$ is the diphoton invariant mass. Scaling the p_T thresholds in this way avoids distortion of the shape of the $m_{\gamma\gamma}$ distribution. The background is estimated from data, without the use of MC simulation, by fitting the diphoton invariant mass distribution in a range ($100 < m_{\gamma\gamma} < 180$ GeV). A polynomial function is used to describe the shape of the background.

The results from the CMS experiments are shown in Fig. 8(a). A clear peak at a diphoton mass of around 125 GeV is seen. A similar result was obtained in the ATLAS experiment.[11]

8.1.2. The $H \to ZZ \to 4l$ decay mode

In the $H \to ZZ \to 4l$ decay mode a search is made for a narrow four-charged lepton mass peak in the presence of a small continuum background. The background sources include an irreducible four-lepton contribution from direct ZZ production via quark-antiquark and gluon-gluon processes. Reducible background contributions arise from $Z_+b\bar{b}$ and $t\bar{t}$ production where

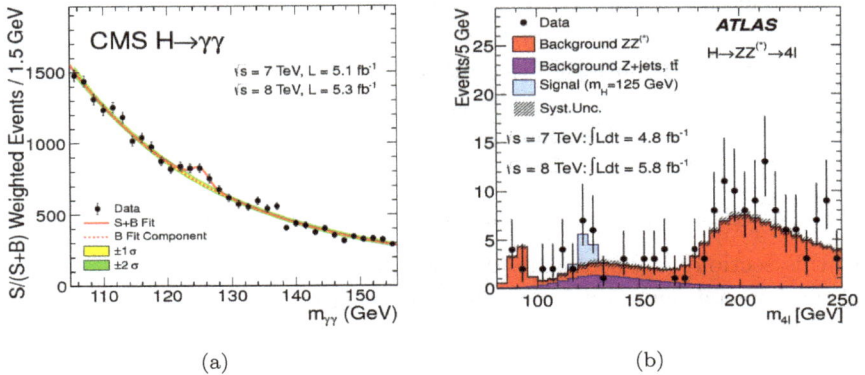

Fig. 8. (a) The two-photon invariant mass distribution of selected candidates in the CMS experiment, weighted by S/B of the category in which it falls. The lines represent the fitted background and the expected signal contribution ($m_H = 125$ GeV). (b) The four-lepton invariant mass distribution in the ATLAS experiment for selected candidates relative to the background expectation. The expected signal contribution ($m_H = 125$ GeV) is also shown.

the final states contain two isolated leptons and two b-quark jets producing secondary leptons.

The event selection requires two pairs of same-flavour, oppositely charged leptons. Since there are differences in the reducible background rates and mass resolutions between the sub-channels $4e$, 4μ, and $2e2\mu$, they are analysed separately. Electrons are typically required to have $p_T > 7$ GeV. The corresponding requirements for muons are $p_T > 5$ GeV. Both electrons and muons are required to be isolated. The pair with invariant mass closest to the Z boson mass is required to have a mass in the range 40–120 GeV and the other pair is required to have a mass in the range 12–120 GeV. The ZZ background, which is dominant, is evaluated from Monte Carlo simulation studies.

The m_{4l} distribution is shown in Fig. 8(b) for the ATLAS experiment.[11] A clear peak is observed at ∼125 GeV in addition to the one at the Z mass. The latter is due to the conversion of an inner bremsstrahlung photon emitted simultaneously with the dilepton pair. A similar result was obtained by the CMS experiment.[10]

8.1.3. *Combinations*

A search was also made in other decay modes of a possible Higgs boson and combined to yield the final results published in August 2012 by ATLAS[10]

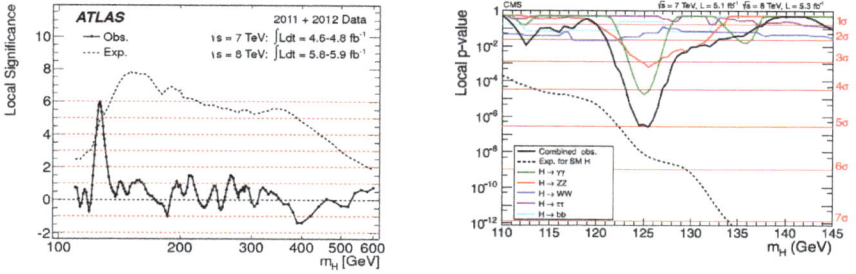

Fig. 9. The combined result of all searches in the ATLAS experiment (left) and the CMS experiment (right) for the observed and expected local significance as a function of mass. Note the vertical axes in the two plots are different.

and CMS[11] experiments. The results (Fig. 9) are presented in terms of local significance for a range of masses. It is clear that both ATLAS and CMS independently discovered a new heavy boson at approximately the same mass, clearly evident in the two different decay modes, $\gamma\gamma$ and ZZ (see Fig. 9(b)). The observed (expected) local significances were 6.0σ (5.0σ) and 5.0σ (5.8σ) in ATLAS and CMS respectively, cleary indicating that a new particle had been discovered.

The decay into two bosons (two γ; two Z bosons; two W bosons) implied that the new particle is a boson with spin different from one and its decay into two photons that it carries either spin-0 or spin-2.

The results presented by both ATLAS and CMS collaborations were consistent, within uncertainties, with the expectations for a SM Higgs boson. Both noted that collection of more data would enable a more rigorous test of this conclusion and an investigation of whether the properties of the new particle imply physics beyond the SM.

8.2. Results from the full 2011 and 2012 data set

Now we present the results from the full dataset corresponding to an integrated luminosity of ~ 5 fb^{-1} at $\sqrt{s} = 7$ TeV and ~ 20 fb^{-1} at $\sqrt{s} = 8$ TeV. This larger dataset also allowed confirmation of the discovery of the new boson, a better examination of the decay channels other than the $H \to \gamma\gamma$ and the $H \to ZZ \to 4l$ decay modes and the first substantial investigations of the boson's properties.

8.2.1. *The $H \to \gamma\gamma$ and the $H \to ZZ \to 4l$ decay modes*

The results from the ATLAS experiment are shown for the $H \to \gamma\gamma$ decay mode (Fig. 10(a))[24] and those from the CMS experiment for the $H \to ZZ \to 4l$ mode (Fig. 10(b)).[25] The signal is unmistakable and the significances have increased as can be seen in Table 3. The data show an even clearer excess of events above the expected background around 125 GeV. The complementary data from the two experiments can be found in Refs. 26 and 27.

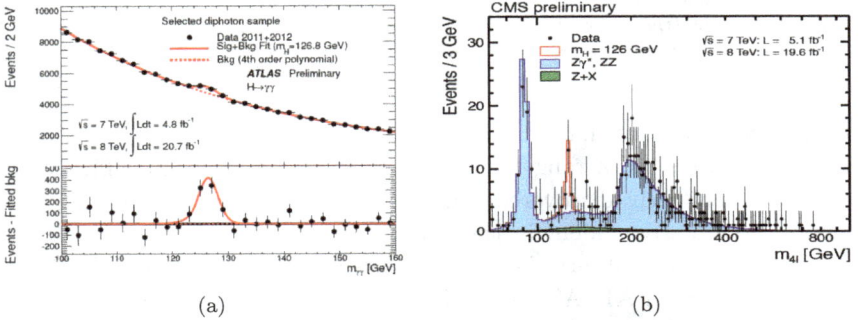

(a)   (b)

Fig. 10.   Invariant mass distribution of di-photon candidates. The result of a fit to the background described by a polynomial and the sum of signal components is superimposed. The bottom inset displays the residuals of the data with respect to the fitted background component. (b) The four-lepton invariant mass distribution in the CMS experiment for selected candidates relative to the background expectation. The expected signal contribution is also shown.

8.2.2. $H \to WW \to 2l\,2\nu$ decay mode

The search for $H \to W^+W^-$ is based on the study of the final state in which both W bosons decay leptonically, resulting in a signature with two isolated, oppositely charged, high p_T leptons (electrons or muons) and large missing transverse momentum, E_T^{miss}, due to the undetected neutrinos. The signal sensitivity is improved by separating events according to lepton flavor; into e^+e^-, $\mu^+\mu^-$, and $e\mu$ samples and according to jet multiplicity into 0-jet and 1-jet samples. The dominant background arises from irreducible non-resonant WW production.

The m_{ll} distribution in the 0-jet in different-flavor final state is shown for CMS in Fig. 11(a).[28] The expected contribution from a SM Higgs boson with $m_H = 125$ GeV is also shown. A background-subtracted distribution of m_T is shown in Fig. 11(b) for the ATLAS experiment.[29] Both show a

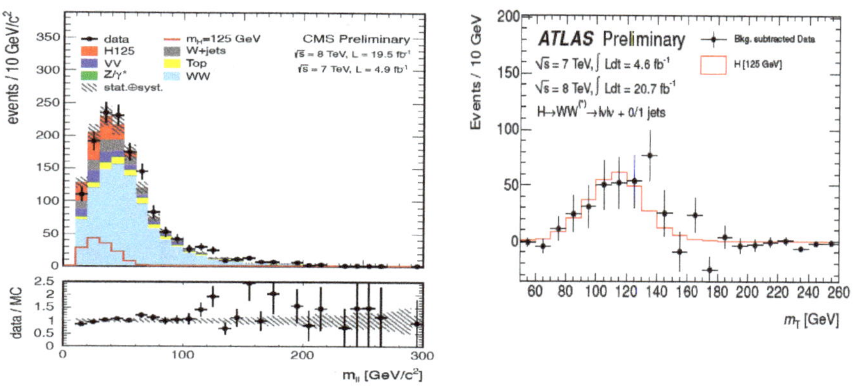

Fig. 11. (a) Distribution in CMS of dilepton mass in the 0-jet in the different-flavor final state for a $m_H = 125$ GeV SM Higgs boson and for the main backgrounds. (b) Background-subtracted m_T distribution. The signal is overlaid.

clear excess of events compatible with a Higgs boson with mass $\sim$125 GeV. The observed (expected) significance of the excess with respect to the background only hypothesis at this mass is 3.9 (5.3) standard deviations in the CMS experiment.[30]

8.2.3. The $H \to \tau\tau$ decay mode

The $H \to \tau\tau$ search is performed using the final-state signatures $e\mu$, $\mu\mu$, $e\tau_h$, $\mu\tau_h$, $\tau_h\tau_h$, where electrons and muons arise from leptonic τ-decays and τ_h denotes a τ lepton decaying hadronically. Each of these categories is further divided into two exclusive sub-categories based on the number and the type of the jets in the event: (i) events with one forward and one backward jet, consistent with the VBF topology, (ii) events with at least one high p_T hadronic jet but not selected in the previous category. In each of these categories, we search for a broad excess in the reconstructed $\tau\tau$ mass distribution. The zero-jet category is used to constrain background normalisation, identification efficiencies, and energy scales. The main irreducible background, $Z \to \tau\tau$ production, and the largest reducible backgrounds (W + jets, multijet production, $Z \to ee$) are evaluated from various control samples in data.

Figure 12 shows the combined observed and expected di-tau mass distributions from CMS,[31] weighting all distributions in each category of each channel by the ratio between the expected signal and background yields for

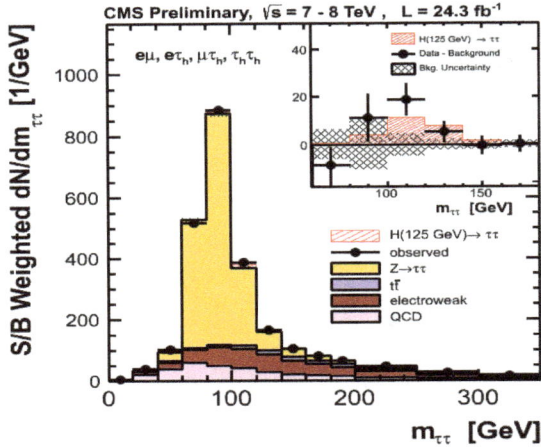

Fig. 12. Combined observed and expected weighted di-tau mass distributions for the various channels. The insert shows the difference between the observed data and expected background distributions, together with the expected signal distribution for a SM Higgs signal at $m_H = 125$ GeV.

the respective category in a di-tau mass interval containing 68% of the signal. A small excess of events is seen around $m_H = 125$ GeV. The plot also shows the difference between the observed data and expected background distributions, together with the expected distribution for a SM Higgs boson signal with $m_H = 125$ GeV. The observed (expected) significance of the excess with respect to the background only hypothesis at this mass is 2.8 (2.6) standard deviations in the CMS experiment.[30] The results include the search for a SM Higgs boson decaying into a τ pair and produced in association with a W or Z boson decaying leptonically.

8.2.4. $H \to b\bar{b}$ decay mode

The $H \to b\bar{b}$ decay mode has by far the largest branching ratio ($\sim 55\%$). However since $\sigma_{bb}(\text{QCD}) \sim 10^7 \times \sigma(H \to b\bar{b})$ the search concentrates on Higgs boson production in association with a W or Z boson using the following decay modes: $W \to e\nu/\mu\nu$ and $Z \to ee/\mu\mu/\nu\nu$. The $Z \to \nu\nu$ decay is identified by the requirement of a large missing transverse energy. The Higgs boson candidate is reconstructed by requiring two b-tagged jets. The search is divided into events where the vector bosons have medium or large transverse momentum and recoil away from the candidate Higgs boson.

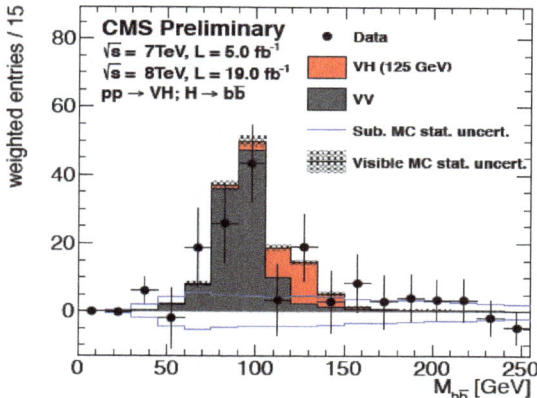

Fig. 13. The weighted dijet invariant mass distribution, combined for all channels, with all backgrounds except dibosons subtracted. The expected signal used corresponds to the production of a Higgs boson with a mass of 125 GeV.

Figure 13 shows the weighted dijet invariant mass distribution in CMS[32] when all backgrounds, except dibosons, are subtracted. For each channel, the relative weight of each $p_T(V)$ bin is obtained from the ratio of the expected number of signal events to the sum of expected signal and background events in a window of $m(jj)$ values between 105 and 150 GeV. The expected signal used corresponds to the production of a Higgs boson with a mass of 125 GeV.

The data are consistent with the presence of a di-boson signal (ZZ and WZ, with $Z \to bb$), together with a small excess consistent with that originating from the production of a 125 GeV SM Higgs boson. For a Higgs boson of mass 125 GeV, the expected and observed 95% CL upper limits on the production cross section, relative to the standard model prediction, are 1.4 and 2.0, respectively.

8.3. Combining the results

8.3.1. Significance of the observed excess

Table 3 summarises in CMS the median expected and observed local significance for a SM Higgs boson mass hypothesis of 125.7 GeV from the individual decay modes.[30] The value of the mass is the measured one, as described in Sec. 6.3.2. The expected significance is evaluated assuming the expected background and signal rates. Figure 14 shows the combination of the data from the various decay modes from the ATLAS experiment

Table 3. The expected and observed local p-values, expressed as the corresponding number of standard deviations of the observed excess from the background only hypothesis, for $m_H = 125.7$ GeV, for various combinations of decay modes.

Decay mode/combination	Expected (σ)	Observed (σ)
$\gamma\gamma$	3.9	3.2
ZZ	7.1	6.9
WW	5.3	3.9
bb	2.2	2.0
$\tau\tau$	2.6	2.8
$\tau\tau + bb$	3.4	3.4

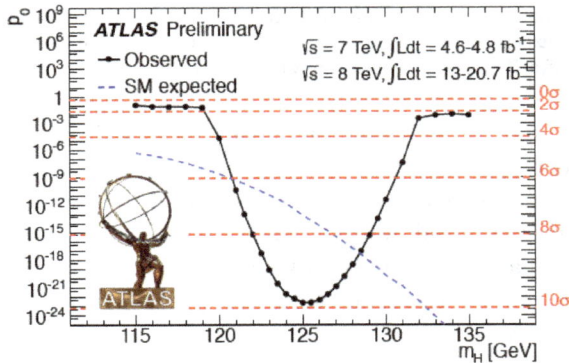

Fig. 14. The ATLAS combination of the various decay modes showing the local probability, p_0, for a background only experiment to be more signal-like than the observation as a function of m_H. The dashed curve shows the median expected local p_0 under the hypothesis of a SM Higgs boson production at that mass.

resulting in an observed significance of $\sim 10\sigma$.[33] Both experiments confirm the observation of a new particle with a mass near 125 GeV.

Both experiments confirm the observation[10,11] of a new particle with a mass near 125 GeV in the $\gamma\gamma$, $ZZ \to 4l$ and $WW \to l\nu l\nu$ channels.

CMS has separately combined the results for the fermionic decay modes ($b\bar{b}$ and $\tau\tau$) and observes a local significance of 3.4σ (Table 3). The Tevatron experiments, CDF and D0, have also reported a combined observed significance of 3.0σ,[34] where the $H \to bb$ mode is the dominant one. Hence existence of the fermionic decays of the new boson, consistent with the expectation from the SM, is established.

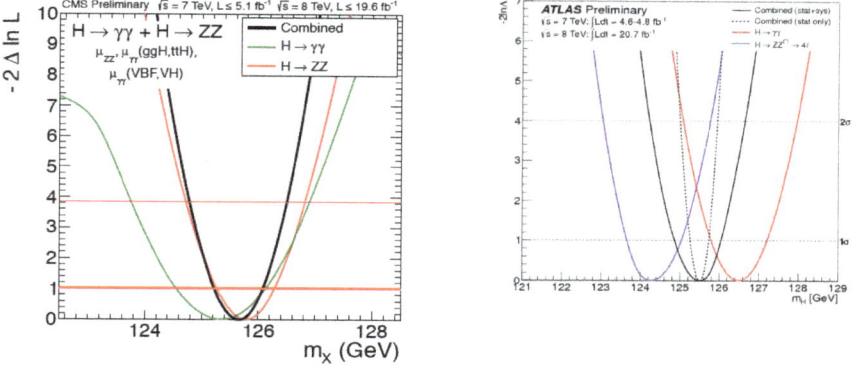

Fig. 15. 1D-scan of the test statistic versus the boson mass m_X or m_H for the $\gamma\gamma$ and $4l$ final states separately and for their combination (left) for CMS and (right) for ATLAS.

8.3.2. *Mass of the observed state*

To measure the mass of the observed state, CMS uses the $ZZ \to 4l$ and $\gamma\gamma$ channels that have excellent mass resolution. To extract the value of mass, m_X, in a model-independent way, the $gg \to H \to \gamma\gamma$, VBF + VH $\to \gamma\gamma$, and $H \to ZZ \to 4l$ processes are assumed to be independent and thus not tied to the SM expectation. The signal in all channels is assumed to be due to a state with a unique mass, m_X. Figure 15 (left) shows the scan of the test statistic as a function of the mass m_X for the two final states separately and their combination in CMS. The intersections of the $q(m_X)$ curves with the horizontal thick line at 1 and thin line at 3.8 define the 68% and 95% CL intervals for the mass of the observed particle, respectively. These intervals include both statistical and systematic uncertainties. The 68% CL interval is $m_X = 125.7 \pm 0.4$ GeV.[30] A similar analysis in the ATLAS experiment yields the results shown in Fig. 15 (right) with $m_X = 125.5 \pm 0.6$ GeV at a CL of 68%.[33]

8.3.3. *Compatibility of the observed state with the SM Higgs boson hypothesis: signal strength*

To establish whether the newly found state is a Higgs boson, or the Higgs boson of the SM, we need to precisely measure its other properties and attributes. The SM Higgs boson is a fundamental scalar particle with spin-parity $J^P = 0^+$, and couples to fundamental fermions as m_f^2/v^2 and to fundamental bosons as m_V^4/v^2 where $v = 246$ GeV. Several tests of

114

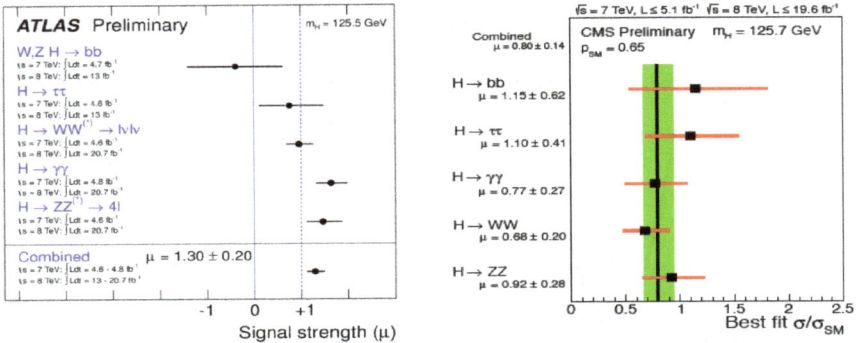

Fig. 16. Values of $\mu = \sigma/\sigma_{\rm SM}$ for sub-combinations by decay mode in (left) ATLAS and (right) in CMS.

compatibility of the observed excesses with those expected from a standard model Higgs boson have been made.

In one comparison labeled as the signal strength μ, the measured production $\times$ decay rate of the signal is compared with the SM expectation, determined for each decay mode individually and for the overall combination of all channels. A signal strength of one would be indicative of a SM Higgs boson.

The best-fit value for the common signal strength μ, obtained in the different sub-combinations and the overall combination of all search channels in the ATLAS and CMS experiments is shown in Fig. 16. The observed μ value is 0.80 ± 0.14 for CMS for a Higgs boson mass of 125.7 GeV[30] and 1.30 ± 0.20 in ATLAS for a Higgs boson mass of 125.5 GeV.[33] In both the experiments the μ-values are consistent with the value expected for the SM Higgs boson ($\mu = 1$). The Tevatron has also measured the value of this signal strength, primarily using the bb channel and find it to be 1.40 ± 0.60.[34]

CMS has also measured μ values, by decay mode and by additional tags used to select preferentially events from a particular production mechanism.[30] Again the results show compatibility with the SM.

8.3.4. Compatibility of the observed state with the SM Higgs boson hypothesis: couplings

The event yield in any (production) $\times$ (decay) mode is usually related to the production cross section, and the partial and total Higgs boson decay

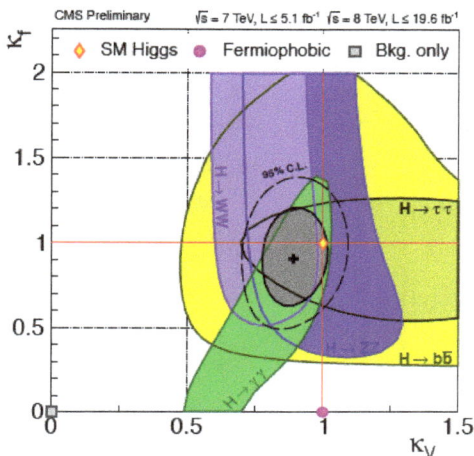

Fig. 17. The 68% CL contours for individual channels (coloured swaths) and the overall combination (solid line) for the (k_V, k_f) parameters. The cross indicates the global best-fit values. The yellow diamond shows the SM point $(k_V, k_f) = (1, 1)$. The result is shown for the positive quadrant only.

widths via the following equation:

$$(\sigma \times \mathrm{BR})(x \to H \to f\!f) = \sigma_x \times \Gamma_{f\!f}/\Gamma_{\mathrm{tot}}$$

where σ_x is the production cross section through the initial state x (x includes gluon-gluon fusion, VBF, WH and ZH, and ttH production mechanisms); $\Gamma_{f\!f}$ is the partial decay width into the final state $f\!f$ (at present f spans WW, ZZ, bb, tt, $\gamma\gamma$, and $Z\gamma$); and Γ_{tot} is the total width of the Higgs boson. The partial widths are proportional to the square of the effective Higgs boson couplings to the corresponding particles. Scale factors, κ_i, modifying the couplings can be introduced to test for possible deviations from the SM expectation. Given the size of the data samples collected so far we only consider $i \equiv V$ for vector bosons or $i \equiv f$ for fermions. Significant deviations of κ from unity would imply new physics beyond the SM Higgs boson hypothesis. The plot of κ_V versus κ_f is shown in Fig. 17 for CMS for results from individual modes and the overall combination.[30] Good compatibility is observed with the SM hypothesis of $\kappa_V = \kappa_f = 1$.

A way to illustrate (Fig. 18) the dependence of the Higgs boson couplings on mass of the decay particles (τ, b-quark, W, Z and t-quark) is to plot these in terms of λ or $\sqrt{(g/2v)}$. For the fermions, the values, λs of the fitted Yukawa couplings $Hf\!f$ are shown, while for vector bosons the square-root of

116

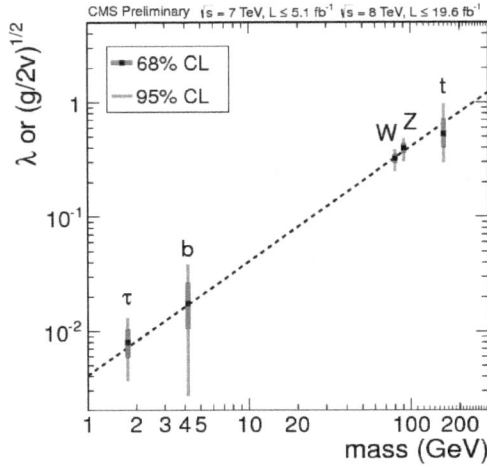

Fig. 18. Summary of the fits from the CMS experiment for deviations in the couplings λ or $\sqrt{(g/2v)}$ as function of particle mass (see text).

the coupling for the HVV vertex divided by twice the vacuum expectation value of the Higgs boson field $\sqrt{(g/2v)}$. The line is the expectation from the SM.

Both Figs. 17 and 18 demonstrate good compatibility with the SM within the errors on the measurements.

8.3.5. Compatibility of the observed state with the SM Higgs boson hypothesis: spin and parity

Key to the identity of the new boson is its quantum numbers amongst which is the spin-parity (J^P). The angular distributions of the decay particles can be used to test various spin hypotheses.

In the decay mode $H \to ZZ^{(*)} \to 4l$ the full final state is reconstructed, including the angular variables sensitive to the spin-parity. The information from the five angles and the two di-lepton pair masses (see Fig. 19) are combined to form boosted decision tree (BDT) discriminants. A decision tree is a set of cuts employed to classify events as "signal-like" or "background-like".

In the decay mode $H \to WW^{(*)} \to l\nu l\nu$ two of the most sensitive variables for measuring spin are the dilepton invariant mass, m_{ll}, and the azimuthal separation of the two leptons, $\Delta\phi_{ll}$.

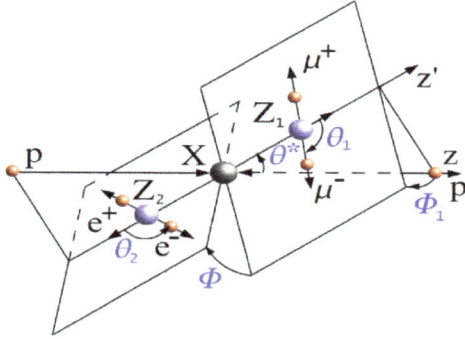

Fig. 19. Illustration of the production and decay of a particle $X \rightarrow Z_1 Z_2 \rightarrow 4l$ with the two production angles θ^* and Φ_1 shown in the X rest frame and three decay angles θ_1, θ_2, and Φ shown in the Z_i and X rest frames, respectively.

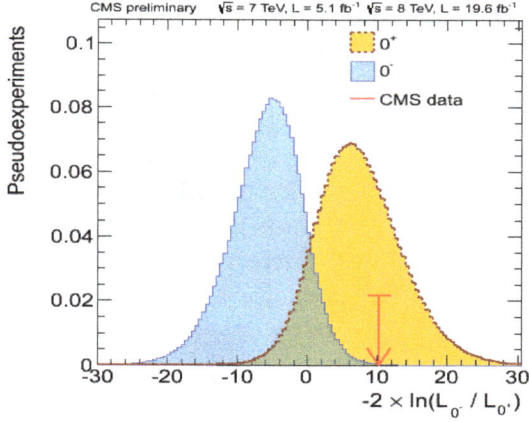

Fig. 20. Distribution of $q = -2\ln(L_{JP}/L_{SM})$ for two signal types, 0^+ represented by the yellow histogram and 0^- hypothesis by the blue histogram for $m_H = 126$ GeV for a large number of generated experiments. The arrow indicates the observed value.

A first study has been presented by CMS in the $ZZ \rightarrow 4l$ channel[35] with the data already disfavouring the pure pseudo-scalar hypothesis (Fig. 20). Further tests have been carried out between the SM Higgs boson with floating signal strength and a scalar ($J^P = 0^+$), a pseudo-scalar (0^-) or a spin-2 resonance with minimal coupling produced in gluon fusion ($2^+_m(gg)$). A more comprehensive study in the same channel has been presented[25]

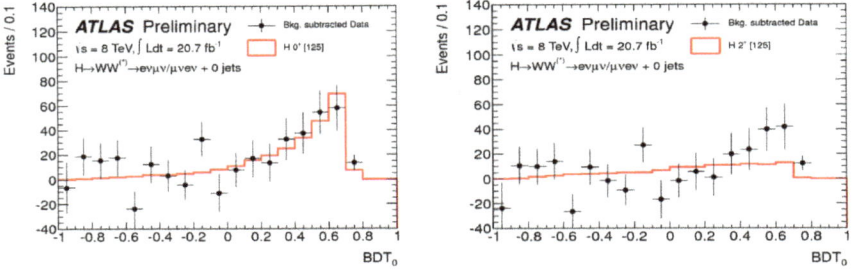

Fig. 21. Background-subtracted data plots of the BDT output distributions using best-fit values for 0^+ (left) and 2_m^+ hypotheses (right). The 2_m^+ hypothesis is shown for $f(qq) = 100\%$.

considering other alternatives ($J^P = 0_h^+$, 0^-, graviton-like tensor with minimal couplings, $2_m^+(gg)$, $2_m^+(qq)$, 1^-, and 1^+ hypotheses. The hypothesis test between 0^+ and $2_m^+(gg)$ has also been carried out in the $H \rightarrow WW \rightarrow l\nu l\nu$ channel.[28]

The CMS has also combined the $ZZ \rightarrow 4l$ and $WW \rightarrow l\nu l\nu$ spin analyses.[30] Under the assumption that the observed boson has $J^P = 0^+$, the data disfavour the hypothesis of a graviton-like boson with minimal couplings produced in gluon fusion, $J^P = 2_m^+$, with a CLs value of 0.60%.

ATLAS has tested the 0^+ and 2_m^+ hypotheses using a two-dimensional kinematic shape fit in the mode $H \rightarrow WW^{(*)} \rightarrow l\nu l\nu$.[36] The discriminants used in the fit are outputs of two different BDTs, trained separately against all backgrounds to identify 0^+ and 2_m^+ events, respectively. For the BDT the kinematic variables used are the transverse mass m_T, $\Delta\phi_{ll}, m_{ll}$ and dilepton p_{ll}^T.

The backgrounds subtracted data are compared (Fig. 21) to the Monte Carlo predictions using the best-fit values for the signal strength for 0^+ and graviton-like tensor 2_m^+ hypotheses. The tested 2_m^+ hypothesis is excluded in favour of a 0^+ hypothesis at a confidence level which varies between 99% for $f(qq) = 100\%$ and 95% for $f(qq) = 0\%$ where $f(qq)$ is the fraction of Higgs boson production via the quark-antiquark annihilation rather than gluon fusion.

ATLAS has also presented a combined study of the spin of the Higgs boson candidate[37] using the $H \rightarrow \gamma\gamma$, $H \rightarrow WW \rightarrow l\nu l\nu$ and $H \rightarrow ZZ \rightarrow 4l$ decays to discriminate between the SM assignment of $J^P = 0^+$ and a specific model of $J^P = 2^+$. The data strongly favour the $J^P = 0^+$ hypothesis. The specific $J^P = 2^+$ hypothesis is excluded with a confidence level above 99.9%, independently of the assumed contributions of gluon fusion and quark-antiquark annihilation processes in the production of the spin-2 particle.

8.3.6. *Compatibility of the observed state with the SM Higgs boson hypothesis: non-standard couplings*

Decays proceeding via quantum loops, such as to two photons, are sensitive to massive but yet undiscovered particles. In such decay modes it is particularly important to see if the measured coupling significantly deviates from the SM expectation. The current results are compatible with those expected from the SM. However the two experiments, although compatible with the SM, measure the signal strength in the two-photon decay modes to be 1.55 ± 0.30 (ATLAS) and 0.77 ± 0.27 (CMS). More data are needed to see if a significant deviation away from $\mu = 1$ will be observed.

9. Conclusions and Outlook

The results from the two experiments show that a scalar Higgs boson has been discovered. It appears to be a state with spin-parity $J^P = 0^+$, with couplings to other SM particles consistent with those predicted for the SM Higgs boson. However, several theories of physics beyond the standard model predict the existence of more than one Higgs boson. One of these would only be subtly different from the SM one. Much more data need to be collected to enable rigorous testing of the compatibility of the new boson with the SM and to establish whether precise measurements of its properties imply the existence of physics beyond the SM.

The LHC accelerator and the experiments have stopped operation and are due to re-commence in the spring of 2015. The LHC accelerator is carrying out maintenance and refurbishment of some parts of the accelerator so that it will be able to reach its full design energy (14 TeV). The experiments are carrying out maintenance, and the installation of the few remaining elements from the baseline design.

It is foreseen that the instantaneous luminosity will be increased by a factor of two and the energy doubled in the next physics run (2015–2018). This should allow an increase, by a factor of around ten, in the integrated data samples. Precise measurements of the properties of the new boson will be made. Furthermore, the energy increase will enable deeper exploration of the physics of the TeV energy scale, especially the physics beyond the SM for which several possibilities are conjectured: supesymmetry, extra

120

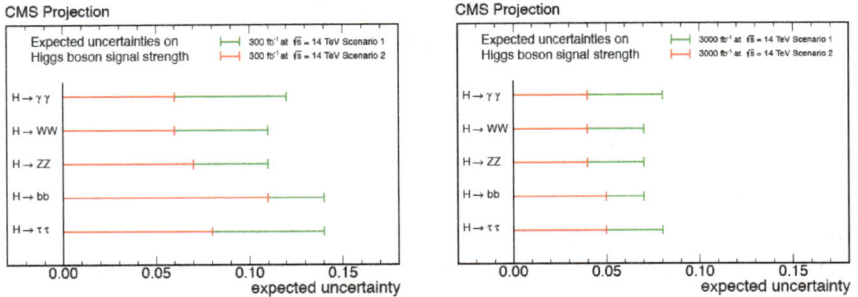

Fig. 22. Estimated precision on the measurements of the signal strength for a SM-like Higgs boson. The projections assume $\sqrt{s} = 14$ TeV and an integrated dataset of 300 fb^{-1} (left) and 3000 fb^{-1} (right). The projections are obtained with the two uncertainty scenarios described in the text.

dimensions, unified theories, superstrings etc. For example, in supersymmetry, five types of Higgs bosons are predicted to exist. Furthermore, the lightest stable neutral particle of this new family of supersymmetric particles could be the particle constituting dark matter. If, as conjectured, such particles are light enough, they ought to reveal themselves at the LHC.

In the longer term it is planned to increase the instantaneous luminosity of the LHC so as to record an integrated luminosity of around 3000 fb^{-1}, by around 2030, instead of the original design value of 300 fb^{-1}. Such an integrated luminosity will require substantial upgrades of the ATLAS and CMS experiments enabling a very precise measurement of the properties of the Higgs boson,[38,39] the study of its rare decay modes and self-coupling, in addition to the search for physics beyond the SM. The projections in Fig. 22[38] scale the errors from current measurements using $1/\sqrt{\text{(integrated luminosity)}}$ for the statistical errors whilst keeping the current theoretical errors (Scenario 1) or halving the theoretical errors (Scenario 2). The extrapolations predict the possibility of measuring the individual signal strengths with a precision of 6–14% for an integrated luminosity of 300 fb^{-1}, and 4–8% for a dataset corresponding to 3000 fb^{-1}. Around 100 million Higgs bosons would have been produced allowing also a search for exotic and rare decays of the new particle.

The discovery of a Higgs boson suggests that we have discovered a fundamental scalar field that pervades our universe. Astronomical and astrophysical measurements point to the following composition of energy-matter in the universe: $\sim 4\%$ normal matter that "shines", $\sim 23\%$ dark matter, and the rest forming "dark energy." Dark matter is weakly and gravitationally interacting matter with no electromagnetic or strong interactions. These are the properties carried by the lightest supersymmetic particle. Hence the question: Is dark matter supersymmetric in nature? Fundamental scalar fields could well have played a critical role in the conjectured inflation of our universe immediately after the Big Bang and in the recently observed accelerating expansion of the universe that, among other measurements, signals the presence of dark energy in our universe.

The discovery of the new boson is widely expected to be a portal to physics beyond the SM. Physicists at the LHC are eagerly looking forward to the higher-energy running of the LHC and to establishing the true nature of the new boson, to find clues or answers to some of the other fundamental open questions in particle physics and cosmology. The physics exploitation of the LHC has just started and the expectations for other discoveries are high over the coming decades.

Epilogue

We are witnessing the "coronation" of the Standard Model of Particle Physics. The Standard Model is a testament to the incredible intellectual insights of its creators and the diligent experimental work of many over the last five decades. Tom Kibble is rightly considered to be one of the creators of the Standard Model.

Tom is well known for his humility which shines through the cover notes, co-signed by Gerry Guralnik and Carl Hagen, that appeared in the special issue of Physics Letters B along with the ATLAS[10] and CMS[11] discovery papers.

"It is somewhat surreal to find that work we did nearly fifty years ago is once again at the centre of attention. This is a triumph for the standard model of particle physics, but even more for the experimenters. The achievement of the two great experimental collaborations reported here is quite magnificent. They have devoted decades to planning, designing, building and operating these huge pieces of precision engineering. It is great to know that the famous boson almost certainly exists, and we are eagerly awaiting for detailed measurement of its properties".

Acknowledgements

The construction, and now the operation and exploitation, of the large and complex ATLAS and CMS experiments have required the talents, the resources, and the dedication of thousands of scientists, engineers and technicians worldwide. This paper is dedicated to all who have worked on these experiments. The superb construction, and efficient operation of the LHC accelerator and the WLCG computing are gratefully acknowledged.

References

1. F. Englert and R. Brout, *Phys. Rev. Lett.* **13** (1964) 321.
2. P. W. Higgs, *Phys. Lett.* **12** (1964) 132.
3. P. W. Higgs, *Phys. Rev. Lett.* **13** (1964) 508.
4. G. S. Guralnik, C. R. Hagen, and T. W. B. Kibble, *Phys. Rev. Lett.* **13** (1964) 585.
5. P. W. Higgs, *Phys. Rev.* **145** (1966) 1156.
6. T. W. B. Kibble, *Phys. Rev.* **155** (1967) 1554.
7. S. L. Glashow, *Nucl. Phys.* **22** (1961) 579.
8. S. Weinberg, *Phys. Rev. Lett.* **19** (1967) 1264.
9. A. Salam, Proceedings of the Eighth Nobel Symposium, ed. N. Svartholm, p. 367 (Almqvist & Wiskell, 1968).
10. ATLAS Collaboration, *Phys. Lett. B* **716** (2012) 1.
11. CMS Collaboration, *Phys. Lett. B* **716** (2012) 30.
12. Frank Close, *The Infinity Puzzle* (Basic Books, 2011).
13. L. Evans (ed.), The Large Hadron Collider, a Marvel of Technology, EPFL Press, 2009; L. Evans and P. Bryant (eds.) and LHC Machine, JINST 03 (2008) S08001.
14. J. R. Ellis, M. K. Gaillard and D. V. Nanopoulos, *Nucl. Phys. B* **106** (1976) 292.
15. ALEPH, CDF, D0, DELPHI, L3, OPAL, SLD Collaborations, the LEP Electroweak Working Group, the Tevatron Electroweak Working Group, and the SLD Electroweak and Heavy Flavour Groups, Precision electroweak measurements and constraints on the standard model, CERN PH-EP-2010-095, http://lepewwg.web.cern.ch/LEPEWWG/plots/winter2012/, arXiv:1012.2367, 2010, http://cdsweb.cern.ch/record/1313716.
16. ALEPH, DELPHI, L3, OPAL Collaborations, and LEP Working Group for Higgs Boson Searches, *Phys. Lett. B* **565** (2003) 61.
17. LHC Higgs Cross Section Working Group, S. Dittmaier, C. Mariotti, G. Passarino and R. Tanaka (Eds.), http://arxiv.org/abs/1101.0593 (2011), http://arxiv.org/abs/1201.3084 (2012).
18. C. J. Seez, T. S. Virdee, L. Di Lella, R. H. Kleiss, Z. Kunszt and W. J. Stirling, in G. Jarlskog and D. Rein (Eds.), Proceedings of the Large Hadron Collider Workshop, Aachen, Germany, 1990, p. 474, CERN 90-10-V-2/ECFA 90-133-Vol-2.
19. M. Della Negra, D. Froidevaux, K. Jakobs, R. Kinnunen, R. Kleiss, A. Nisati and T. Sjostrand, in G. Jarlskog and D. Rein (Eds.), Proceedings of the Large Hadron Collider Workshop, Aachen, Germany, 1990, p. 509, CERN 90-10-V-2/ECFA 90-133-Vol-2.
20. N. Ellis and T. S. Virdee, *Ann. Rev. Nucl. Sci.* **44** (1994) 609.
21. CMS Collaboration, Letter of Intent, CERN-LHCC-92-003 (1992); Technical Proposal, CERN-LHCC-1994-038 (1994); JINST 3 (2008) S08004.

22. ATLAS Collaboration, ATLAS: Letter of Intent, CERN-LHCC-92-004 (1992); Technical Proposal, CERN-LHCC-1994-043 (1994); JINST 3 (2008) S08003.

23. The LHC Detector Challenge, *Physics World*, Vol. 17, No. 9, (2004); Detectors at LHC, *Phys. Rep.* **403–404** (2004) 401.

24. ATLAS Collaboration, Measurements of the properties of the Higgs-like boson in the two-photon decay channel with the ATLAS detector using 25 fb^{-1} of proton-proton collision data, ATLAS-CONF-2013-012 (2013).

25. CMS Collaboration, Properties of the Higgs-like boson in the decay $H \to ZZ \to 4l$ in pp collisions at $\sqrt{s} = 7$ and 8 TeV, CMS PAS HIG-13-002 (2013).

26. ATLAS Collaboration, Measurements of the properties of the Higgs-like boson in the four-lepton decay channel with the ATLAS detector using 25 fb^{-1} of proton-proton collision data, ATLAS-CONF-2013-013 (2013).

27. CMS Collaboration, Updated measurements of the Higgs boson at 125 GeV in the two photon decay channel, CMS PAS HIG-13-001 (2013).

28. CMS Collaboration, Update on the search for the standard model Higgs boson in pp collisions at the LHC decaying to W^+W^- in the fully leptonic final state, CMS PAS HIG-13-003 (2013).

29. ATLAS Collaboration, Measurements of the properties of the Higgs-like boson in the $WW^{(*)} \to l\nu l\nu$ decay channel with the ATLAS detector using 25 fb^{-1} of proton-proton collision data, ATLAS-CONF-2013-030 (2013).

30. CMS Collaboration, Measurements of the properties of the new boson with a mass near 125 GeV, CMS PAS HIG-13-005 (2013).

31. CMS Collaboration, Search for the Standard-Model Higgs boson decaying to tau pairs in proton-proton collisions at $\sqrt{s} = 7$ and 8 TeV, CMS PAS HIG-13-004 (2013).

32. CMS Collaboration, Search for the standard model Higgs boson produced in association with W or Z bosons, and decaying to bottom quarks, CMS PAS HIG-13-012 (2013).

33. ATLAS Collaboration, Combined coupling measurements of the Higgs-like boson with the ATLAS detector using up to 25 fb^{-1} of proton-proton collision data. ATLAS-CONF-2013-034 (2013).

34. CDF and D0 Collaborations, Higgs boson studies at the Tevatron, http://arxiv.org/abs/1303.6346 (2013).

35. CMS Collaboration, Study of the mass and spin-parity of the Higgs boson candidate via its decays to Z boson pairs, Phys. Rev. Lett. **110** (2013) 081803.

36. ATLAS Collaboration, Study of the spin properties of the Higgs-like boson in the $H \to WW^{(*)} \to e\nu\mu\nu$ channel with 21fb^{-1} of $\sqrt{s} = 8$ TeV data collected with the ATLAS detector, ATLAS-CONF-2013-031 (2013).

37. ATLAS Collaboration, Study of the spin of the new boson with up to 25 fb^{-1} of ATLAS data, ATLAS-CONF-2013-040 (2013).

38. CMS Collaboration, Projected Performance of an Upgraded CMS Detector at the LHC and HL-LHC, CMS Note-13-002, arXiv:1307.7135.

39. ATLAS Collaboration, Physics at a High-Luminosity LHC with ATLAS, ATL-PHYS-PUB-2013-007, arXiv:1307.7292.

TOM KIBBLE: BREAKING GROUND AND BREAKING SYMMETRIES

STEVEN WEINBERG

Department of Physics, University of Texas at Austin, USA

I'm very pleased to have a chance to help say Happy Birthday to my old friend Tom Kibble. I first met Tom a little over 50 years ago, when I was spending a year here at Imperial College, in the theory group then headed by Abdus Salam. I already knew about some work that Tom had done on a new approach to the general theory of relativity, which I thought was very interesting and attractive. I believe that we spent some time talking about that work, but that year I was preoccupied with a different matter: the spontaneous breakdown of symmetry principles. I was running into a problem, one that Tom and his colleagues were to solve a couple of years later. Their solution played an essential part in the eventual unification of the weak and electromagnetic forces, and the formulation of what is now called the Standard Model of elementary particles, including the new particle discovered last year at CERN.

The particle physicists here will know what I am talking about, but for others I had better say a bit about what symmetry principles are, what it means for them to be broken, and what this has to do with the real world.

First, a symmetry principle is a statement that something looks the same from different points of view. In other words, it is a principle of *invariance*. This can be as simple as the symmetry of a butterfly, for which the symmetry principle says that its appearance doesn't change when you look at it in a mirror, which interchanges right and left. But symmetries can be much more complicated. A square looks the same when you rotate your point of view by any multiple of 90 degrees, and a circle looks the same when you rotate your point of view by any angle. By the way, tonight I will be speaking only of symmetry operations that can be continuously varied, like the rotations of a circle, unlike the discrete symmetries of a butterfly or a square.

All very pretty, but what really interests us physicists is not the symmetries of things like butterflies or squares or circles, but the symmetries of the laws of nature. These are statements that the laws of nature we discover don't change when we change our point of view in certain specific ways.

Some of these symmetries have to do with changes in our orientation in space and time. For instance, Einstein's Special Theory of Relativity is based on the principle, known as Lorentz invariance, that the laws of nature take the same form when we view nature from any uniformly moving laboratory.

Other symmetries have to do with the nature of the particles of matter. In the 1930s it was discovered that there was an approximate symmetry between protons and neutrons, the two types of particle that make up atomic nuclei. It was discovered that the equations governing nuclear forces take the same form not only if we interchange protons and neutrons in these equations, but even if we change protons and neutrons in the equations into particles that are part way between being a proton and being a neutron.

Why is this important? If in the 1930s we had known the laws governing nuclear forces, then it wouldn't have been so important to notice that the laws had this symmetry. But we didn't know these laws. The symmetry was just suggested by similarities in the scattering of protons or neutrons by protons. Still we could use the symmetry of the laws to draw the conclusion that corresponding states of different nuclei form families — not just the simple family consisting of the neutron and proton, but families of complex nuclei as well. Verifying these predictions confirmed the symmetry, and we learned a lot about atomic nuclei, even without understanding anything else about nuclear forces. More important, we could be confident in using the symmetry as a clue that would eventually help us to find the true theory of the strong nuclear force.

In 1960 some theorists fell in love with a new idea that grew out of the work of Yoichiro Nambu, which in turn grew out of experience with superconductors and iron magnets. Symmetries may be broken — that is, they can be obeyed (perhaps exactly) by the equations of physics, and yet not obeyed even approximately by the solutions of these equations, the observable phenomena of nature. So we suddenly realized that the laws of nature may obey many more symmetries that were still to be discovered, symmetries that weren't already apparent in the family relationships of known particles. It's a very Platonic idea, that at the deepest level nature has a perfection that is not readily apparent. Some of us theoretical physicists felt like children, finding a previously locked cupboard, filled with lovely jars of jam.

Almost immediately when we tried to open this cupboard, an alarm sounded. Broken symmetries have consequences, not necessarily desirable. Jeffrey Goldstone, who was then a research fellow at Cambridge, found in various examples (as Nambu had already found in the theory he had studied) that whenever an exact symmetry of the equations of physics is broken, there must appear a particle of zero mass and zero spin. These came to be called Goldstone particles, or Nambu-Goldstone particles.

But no such particle was known in nature. Now, it's not uncommon for theorists to speculate about particles that have not yet been observed, because if we don't know anything about the mass of these particles, we can always suppose that they are just too heavy to have been produced at existing accelerators. But that excuse couldn't be used with a particle that was predicted to have zero mass — such particles would have been coming out of the ears of experimental physicists.

While I was here at Imperial College I collaborated with Salam, and at long distance with Goldstone, to turn the anecdotal appearance of Goldstone particles into a theorem, which would regrettably kill off the idea of broken symmetries. We showed with apparent mathematical rigor that whenever an exact symmetry is broken, there must appear in nature a corresponding massless spinless particle. This was terribly disappointing — it seemed to spell the end of search for new symmetries in nature.

This brings me back to Tom Kibble. In 1964, together with two visitors to Imperial College, Gerald Guralnik and Carl Hagen, Tom pointed out an exception to the theorem of Goldstone, Salam, and myself: The theorem doesn't apply in theories like electrodynamics, in which the symmetry principle in question is associated with a field like the electromagnetic field. (In electrodynamics the symmetry is known as gauge invariance, which says among other things that only differences in voltage matter.) Instead of a Goldstone particle appearing in the theory, the particles similar to photons associated with this field acquire a mass. (A similar conclusion was reached independently at about the same time by Robert Brout and François Englert in Brussels and by Peter Higgs in Edinburgh.) The reason, as Kibble and his co-workers were kind enough to point out, is not that Goldstone, Salam, and I had made a mathematical mistake, but rather that we had relied on the Einstein symmetry, Lorentz invariance, that I mentioned earlier, which says that the laws of nature look the same to all uniformly moving observers. Although this symmetry does apply to anything that is actually observable, it doesn't apply to the way that electromagnetism is treated mathematically in quantum mechanics. This was not just a technicality — the door was now open to exploring theories with broken symmetries, without having to worry about Goldstone particles.

But it took a few years to go through this door. It wasn't until 1967 that Kibble presented an important generalization of his work with Guralnik and Hagen, which brought this work closer to an application to the real world. He showed that the all-dreaded Goldstone particles were absent in a much larger class of theories, all those in which the symmetry is *local*, in the sense that the symmetry transformations, that according to the symmetry do not change the laws of nature, are allowed to vary from point to point in space and time. Such a local theory had been proposed in 1954 by Chen-Ning Yang and Robert Mills, in an attempt to account for the strong nuclear forces that hold protons and neutrons together in atomic nuclei. (Also, Kibble's version of general relativity, which I mentioned earlier, was based on making Lorentz invariance into a local symmetry.) Physicists had pretty well given up on using these Yang-Mills theories as a basis for physical theories, because they seemed to require that the particles that transmit the force would have zero mass, like the photon and graviton. These would not be Goldstone particles, but would have a spin, equal to one in natural units, the same spin as the photon. Of course, no such particles were known, aside from the photon itself. But Kibble now showed that, as in the simpler theories studied in 1964, if the local symmetry is broken, then these force-carrying particles would acquire a mass, and so if heavy enough could have escaped detection.

Even so, no specific physical theory was being proposed. One reason that this took a while is that, starting around 1965, theorists began to make serious use of an idea going back to Nambu, that a known particle, the pion, is a Goldstone particle — not the massless Goldstone particle expected for a broken exact symmetry, but a relatively light particle associated with a broken *approximate* symmetry, a generalization of the proton-neutron symmetry I mentioned earlier. My own work from 1965 to 1967 centered on this idea. The Goldstone particles of approximate symmetries are known as pseudo-Goldstone particles, a term I invented, and that Goldstone hated. So because we were recognizing pseudo-Goldstone particles in nature, we weren't so keen any more on getting rid of Goldstone particles.

In 1967 I turned away from pions and strong nuclear forces, and started to think about using what I had learned from that work to construct a mathematically satisfactory theory of the weak nuclear forces. These are the forces that, instead of holding neutrons and protons together in atomic nuclei, allow protons to turn into neutrons and vice versa, with the emission of electrons and neutrinos, reactions that occur in a kind of radioactivity, and that also fuel the Sun. It was an old idea that these forces are transmitted by some sort of heavy charged particle, called the W (for "weak", not Weinberg), with the same spin as the photon. Maybe these force-carrying particles got their mass in the way described by Kibble, from the breaking of a local symmetry. The simplest way of implementing this idea would require two other force-carrying particles, another heavy one that I called the Z, and a massless particle, which clearly could be identified as the photon. (The same theory was developed independently by Salam.) Why is the photon massless? The reason was that the symmetry of gauge invariance, with which the photon is associated, is not broken, and as Kibble had pointed out, this meant that it would remain massless while all the other force-carrying particles were getting heavy. Experiments in the 1970s and 1980s showed that this is indeed the correct theory of weak and electromagnetic forces.

By the way, Goldstone particles may not be entirely dead, provided they are associated with the breakdown of symmetries that are *not* local. The cosmic background of microwave radiation is somewhat sensitive to the number of species of neutrinos that were present in the early universe. Measurements from the WMAP satellite (and now the Planck satellite) have persistently given a value for this number between 3 and 4, though only 3 species are known. A massless Goldstone particle associated with the breakdown of symmetries of dark matter could look very much like 4/7 (or perhaps somewhat less) of a neutrino species. This is just a wild speculation, but if it turns out to be correct, please remember that you heard it here first.

I need to bring up one more point about broken symmetries, that has been much in the news lately. Broken symmetries don't get broken by themselves. For instance, the laws that govern the atoms in a bar of iron and the forces between them respect a perfect symmetry between different directions, but this symmetry can be broken by the spontaneous appearance of a magnetic field, when the iron is cooled below a temperature of about 770 degrees Centigrade. The magnetic field must point in some direction, and it causes the spins of all the iron atoms to line up in that direction. Similarly, when Goldstone and those after him explored the idea of broken symmetry, they introduced one or more fields to break the symmetry. Since no one wants to break the symmetry between directions in space, the fields introduced in these theories are unlike a magnetic field; they do not have a sense of direction in space. Such fields are called *scalar*. These scalar fields instead distinguish between different types of particle with which they interact, and when they spontaneously turn on as the universe cools they break various symmetries among these particles. Such fields were introduced by Goldstone in 1960, and in the theories studied in 1964 by Guralnik, Hagen, and Kibble as well as by Brout and Englert, and Higgs.

In all these theories the various scalar fields show up in two different ways. Some of them appear as Goldstone particles, which for local symmetries aren't real particles, but as Kibble and the others showed just serve to give mass to the photon-like force-carrying particles associated with these symmetries. But others show up as real physical particles of zero spin (because the fields are scalar) and non-zero mass (because they aren't Goldstone particles). We needed a name for the latter massive particles, and although they had already appeared in Goldstone's 1960 work, we could hardly call them Goldstone particles, so they came to be called Higgs particles, though they could just as well have been called Brout-Englert-Higgs-Guralnik-Hagen-Kibble particles, or just Kibble particles.

For an example, think of a circular valley, with perfect circular symmetry — every cross section through the valley looks the same as every other. Drop a ball into the valley; wherever it comes to rest on the valley floor breaks this symmetry. But the symmetry is still there; it tells us that the valley floor is perfectly flat, so an infinitesimal perturbation can get the ball rolling all the way around the valley floor. This mode of motion corresponds to a Goldstone boson, which because it has zero mass can be created with any tiny energy. But given enough energy, the ball can also be made to roll up and down the sides of the valley. This mode corresponds to a Higgs particle.

No specific prediction about Higgs particles were made by Goldstone or in the 1964 papers, because no specific theories were being proposed. When the theory of weak and electromagnetic forces was introduced in 1967, a Higgs particle appeared in the theory, all of whose properties were predicted by the theory, except for the particle's mass (and name). The Large Hadron Collider at CERN was designed in part to be able to produce Higgs particles, and as you probably know, a particle was found last year that so far appears to be the predicted Higgs particle, putting in place the final piece of the Standard Model of elementary particles.

Now I want to come back to Tom Kibble. I have been talking about his work on broken local symmetries, because that's what I know best, but he's done many other things. Just to mention one of them, as discussed this morning by Neil Turok, Kibble founded the study of cosmic strings and other discontinuities in space that in some theories appear spontaneously as the universe expands. This is a very different example of a broken symmetry. As Kibble showed, knowing what symmetry is broken, and what smaller symmetry survives the breakdown, tells you immediately what kind of stable discontinuities can form: sheets, or vortex lines, or point knots like monopoles. This has become a subject of active experimental research on fluids here and earth, and a continuing concern of astrophysical theorists. Tom is a wonderful theoretical physicist.

I have happy memories of my year at Imperial College — I got to know London a little, I got to know Abdus Salam a little, and not least of my happy memories is that I became friends with Tom Kibble.

TOM KIBBLE AT 80: AFTER DINNER SPEECH

FRANK CLOSE

Department of Physics, University of Oxford, UK

It's easier being Beethoven or Shakespeare than a theoretical physicist.

Change a few words in Hamlet, or some notes in a symphony, and you still have wonderful works of art. But change one symbol in the equations for the mass mechanism, and nothing works at all. However beautiful your theoretical construction, the hard reality of theoretical physics is: experiment decides.

Many here will have seen experiments bin their best ideas faster than the speed of thought. Tom — you are one of the fortunate few who have known the great mystery, the weird ability of maths to know Nature. For the massive boson named after Higgs, whose existence and properties confirm the theory, has been discovered — I think we can say as the lawyers do: "Beyond reasonable doubt". And it has taken less than 50 years! (String theorists would probably settle for that).

It's sobering to realise it's 50 years since the Gang of Six had their breakthrough.

Imagine turning the clock back to 1964; and then looking back from that moment a similar span. It brings us to Rutherford's nucleus and Bohr's model of the atom — whose centenary is this year. It is these vast spans of time – halfway to the birth of modern atomic physics, an entire professional life as a physicist — that are almost as awesome as the achievements themselves.

The ideas of 1964 described how, courtesy of the weak force, the sun shines slowly enough for life to have evolved. They show why there is structure — the extension to fermions explaining why atoms have size (due to the electron mass being non zero) and the nucleus is compact (thanks to non zero quark masses spoiling chiral symmetry) — though this has yet to be proved.

134

You have explained the foundations of chemistry — Rutherford would have approved.

1964 at Imperial College was singular year. Tom, Dick (Hagen), and Gerry (Guralnik) were discovering the mass mechanism while in the same corridors, within weeks, Abdus Salam and John Ward had discovered the key to uniting weak and electromagnetic interactions in $SU(2) \times U(1)$. While researching Infinity Puzzle[1] I never understood, and still don't, why these two seminal ideas were not united there and then. Gerry has told of his lunch experience with John Ward, where each had one half of the jigsaw but failed to complete the conversation that would have brought the pieces together and completed the picture. Instead, 3 years elapsed before Tom took the step which, it seems to me, touched the real world and set in train the real saga: he brought group theory into the mechanism in such a way that what we now call the W and Z become massive while the photon stays massless. In so doing, he showed how it is possible to marry the two forces, and to avoid the paradox of a universe filled with ether: Michelson and Morley used a probe that is impervious to the stuff!

Why do I single out Tom here?

First, he inspired Abdus Salam. Abdus was a group theory person; it was his bread and butter. He realised that Tom had the key, which completed his and Ward's $SU(2) \times U(1)$ model, leading to his half of the Weinberg Salam model. Steve in his seminal paper has cited as his reference number 4 the key paper of Tom, which fused the pieces for him too. This also explains the enigmatic reference of Abdus Salam to the "Higgs Kibble mechanism". It took six people, at least, to make the W and Z bosons massive; it took Tom alone to show how to keep the photon light.

October 1967 was also singular for me as I started as a graduate student in Oxford. That must be the reason why I was asked to speak tonight: it is hard to find people of a generation that both lived through Haight Ashbury and can still remember it. But I cannot remember any of this! At Oxford we met Regge poles; current algebra; quark model — nothing of spontaneous symmetry breaking and the mass mechanism. There was a deafening silence — everywhere.

Unknown to me, in Utrecht Gerard 't Hooft was also starting his career, and showed that the mass mechanism was key to the viable theory — quantum flavourdynamics — which he and Tini Veltman completed, and which has passed all the tests for 40 years.

Since that time, first the W and Z were found; having them in the bag, quantum precision led to predictions of the top quark, whose discovery and further quantum precision heralded the Higgs boson — or as Peter modestly says, the boson named after him.

Having found it, precision measurements of its decays to pairs of photons might give hints of physics yet to be discovered. Why the neutron is heavier than the proton, and why an electron's mass fits so conveniently that beta decays and transmutations occur: these are questions for another day. I hope that finding answers doesn't take another 49 years.

References

1. Frank Close, *The Infinity Puzzle* (Oxford University Press, Basic Books, New York, 2013).

Publication List — Tom W.B. Kibble

Research papers in refereed journals (only published or accepted papers are numbered)

1. T.K. and J.C. Polkinghorne, On Schwinger's variational principle, *Proc. Roy. Soc.* **A242**, 252–263 (1957).
2. Dispersion relations for inelastic scattering, *Proc. Roy. Soc.* **A244**, 355–376 (1958).
3. T.K. and J.C. Polkinghorne, Higher-order spinor Lagrangians, *Il Nuovo Cimento, ser. X* **8**, 74–83 (1958).
4. On the consistency of Schwinger's action principle, *Il Nuovo Cimento, ser. X*, **10**, 417–427 (1958).
5. The commutation relations obtained from Schwinger's action principle, *Proc. Roy. Soc.* **A249**, 441–444 (1959).
6. J.J. Giambiagi and T.K., Jost functions and dispersion relations, *Annals of Physics (N.Y.)* **7**, 39–51 (1959).
7. Kinematics of general scattering processes and the Mandelstam representation, *Phys. Rev.* **117**, 1159–1161 (1960) [doi:10.1103/PhysRev.117.1159].
8. Lorentz invariance and the gravitational field, *J. Math. Phys.* **2**, 212–221 (1961).
9. Feynman rules for Regge particles, *Phys. Rev.* **131**, 2282–2291 (1963) [doi:10.1103/PhysRev.131.2282].
10. Canonical variables for the interacting gravitational and Dirac fields, *J. Math. Phys.* **4**, 1433–1437 (1963).
11. L.S. Brown and T.K., Interaction of intense laser beams with electrons, *Phys. Rev.* **133**, A705–719 (1964) [doi:10.1103/PhysRev.133.A705].
12. G.S. Guralnik, C.R. Hagen and T.K., Global conservation laws and massless particles, *Phys. Rev. Letters* **13**, 585–587 (1964) [doi:10.1103/PhysRevLett.13.585].

13. Frequency shift in high-intensity Compton scattering, *Phys. Rev.* **138**, B740–753 (1965) [doi:10.1103/PhysRev.138.B740].

14. Conservation laws for free fields, *J. Math. Phys.* **6**, 1022–1026 (1965).

15. G.S. Guralnik and T.K., Lagrangian formalism of $\tilde{U}(12)$ symmetry and the Bargmann–Wigner equations, *Phys. Rev.* **139**, B712–719 (1965).

16. Relativistic transformation laws for thermodynamic variables, *Il Nuovo Cimento, ser. X* **41B**, 72–78 (1966).

17. Comment on the remarks of Gamba, *Il Nuovo Cimento, ser. X* **41B**, 83 (1966).

18. Remarks on the comments of Dr Arzeliès, *Il Nuovo Cimento, ser. X,* **41B**, 84–85 (1966).

19. Relativistic corrections to Thomson scattering from laser beams, *Physics Letters* **20**, 627–628 (1966).

20. Refraction of electrons by intense electromagnetic waves, *Phys. Rev. Letters* **16**, 1054–1056 (1966) [doi:10.1103/PhysRevLett.16.1054].

21. Mutual refraction of electrons and photons, *Phys. Rev.* **150**, 1060–1069 (1966) [doi:10.1103/PhysRev.150.1060].

22. Symmetry-breaking in non-Abelian gauge theories, *Phys. Rev.* **155**, 1554–1561 (1967) [doi:10.1103/PhysRev.155.1554].

23. Coherent soft-photon states and infrared divergences. I. Classical currents, *J. Math. Phys.* **9**, 315–324 (1968).

24. Coherent soft-photon states and infrared divergences. II. Mass-shell singularities of Green's functions, *Phys. Rev.* **173**, 1527–1535 (1968) [doi:10.1103/PhysRev.173.1527].

25. Coherent soft-photon states and infrared divergences. III. Asymptotic states and reduction formulas, *Phys. Rev.* **174**, 1882–1901 (1968) [doi:10.1103/PhysRev.174.1882].

26. Coherent soft-photon states and infrared divergences. IV. The scattering operator, *Phys. Rev.* **175**, 1624–1640 (1968) [doi:10.1103/PhysRev.175.1624].

—. Theory of measurements in finite space-time regions, unpublished (1970).

—. Are there superselection rules? Unpublished (1972).

27. T.K., A. Salam and J. Strathdee, Intensity-dependent mass shift and symmetry breaking, *Nucl. Phys.* **B96**, 255–263 (1975) [doi:10.1016/0550-3213(75)90581-7].

28. Topology of cosmic domains and strings, *J. Phys. A: Math. & Gen.* **9**, 1387–1398 (1976) [doi:10.1088/0305-4470/9/8/029].

29. Relativistic models of non-linear quantum mechanics, *Commun. Math. Phys.* **64**, 73–82 (1978) [doi:10.1007/BF01940762].
30. Geometrization of quantum mechanics, *Commun. Math. Phys.* **65**, 189–201 (1979) [doi:10.1007/BF01225149].
31. T.K. and S. Randjbar-Daemi, Non-linear coupling of quantum theory and classical gravity, *J. Phys. A: Math. & Gen.* **13**, 141–148 (1980) [doi:10.1088/0305-4470/13/1/015].
32. S. Randjbar-Daemi, B.S. Kay and T.K., Renormalization of semi-classical theories, *Physics Letters* **91B**, 417–420 (1980), Imperial/TP/79-80/12 [doi:10.1016/0370-2693(80)91010-2].
33. T.K., G. Lazarides and Q. Shafi, Strings in SO(10), *Physics Letters* **113B**, 237–139 (1982), Imperial/TP/81-82/18 [doi:10.1016/0370-2693(82)90829-2].
34. T.K., G. Lazarides and Q. Shafi, Walls bounded by strings, *Phys. Rev. D* **26**, 435–439 (1982) [doi:10.1103/PhysRevD.26.435].
35. T.K. and N. Turok, Self-intersection of cosmic strings, *Physics Letters* **116B**, 141–143 (1982), Imperial/TP/81-82/27 [doi:10.1016/0370-2693(82)90993-5].
36. P. Bhattacharjee, T.K. and N. Turok, Baryon number from collapsing cosmic strings, *Physics Letters* **119B**, 95–96 (1982), Imperial/TP/81-82/31 [doi:10.1016/0370-2693(82)90252-0].
37. M. Hindmarsh and T.K., Monopoles on strings, *Phys. Rev. Letters* **55**, 2398–2400 (1985), Imperial/TP/85-86/01 [doi:10.1103/PhysRevLett.55.2398].
38. M. Hindmarsh and T.K., Hindmarsh and Kibble respond, *Phys. Rev. Letters* **57**, 647 (1986).
39. Configuration of Z_2 strings, *Physics Letters* **B166**, 311–313 (1986), Imperial/TP/84-85/39 [doi:10.1016/0370-2693(86)90806-3].
40. String-dominated universe, *Phys. Rev. D* **33**, 328–332 (1986), Imperial/TP/84-85/30 [doi:10.1103/PhysRevD.33.328].
41. E. Copeland, D. Haws, T.K., D. Mitchell and N. Turok, Monopoles connected by strings and the monopole problem, *Nucl. Phys.* **B298**, 445–447 (1988), Imperial/TP/86-87/24 [doi:10.1016/0550-3213(88)90350-1].
42. T.K. and E. Weinberg, When does causality constrain the monopole abundance? *Phys. Rev. D* **43**, 3188–3190 (1991), Imperial/TP/89-90/33 [doi:10.1103/PhysRevD.43.3188].
43. E. Copeland, T.K. and D. Austin, Scaling solutions in cosmic string networks, *Phys. Rev. D* **45**, R1000–1004 (1992), Imperial/TP/90-

91/19 [doi:10.1103/PhysRevD.45.1000].

44. R. Holman, T.K. and S.-J. Rey, How efficient is the Langacker-Pi mechanism of monopole annihilation? *Phys. Rev. Letters* **69**, 241–244 (1992), Imperial/TP/91-92/18, hep-ph/9203209 [doi:10.1103/PhysRevLett.69.241].

45. D. Austin, E. Copeland and T.K., Evolution of cosmic string configurations, *Phys. Rev. D* **48**, 5594–5627 (1993), Imperial/TP/92-93/42, hep-ph/9307325 [doi:10.1103/PhysRevD.48.5594].

46. A. Gill and T.K., Cosmic rays from cosmic strings, *Phys. Rev. D* **50**, 3660–3665 (1994), Imperial/TP/93-94/23, hep-ph/9403395 [doi:10.1103/PhysRevD.50.3660].

47. L.M.A. Bettencourt and T.K., Non-intercommuting configurations in the collisions of type I U(1) cosmic strings, *Physics Letters* **B332**, 297–304 (1994), Imperial/TP/93-94/34, hep-ph/9405221 [doi:10.1016/0370-2693(94)91257-2].

48. X.A. Siemens and T.K., High-harmonic configurations of cosmic strings: an analysis of self-intersections, *Nuclear Physics* **B438**, 307–319 (1995), Imperial/TP/94-95/01, hep-ph/9412216 [doi:10.1016/0550-3213(94)00592-3].

49. D. Austin, E.J. Copeland and T.K., Characteristics of cosmic-string scaling configurations, *Phys. Rev. D* **51**, R2499–2503 (1995), Imperial/TP/93-94/45, hep-ph/9406379 [doi:10.1103/PhysRevD.51.2499].

—. T.K. and A. Vilenkin, Density of strings formed at a second-order cosmological phase transition, Imperial/TP/94-95/09A, hep-ph/9501207, submitted to *Phys. Rev. Letters* (withdrawn).

50. T.K. and A. Vilenkin, Phase equilibration in bubble collisions, *Phys. Rev. D* **52**, 679–688 (1995), Imperial/TP/94-95/11, hep-ph/9501266 [doi:10.1103/PhysRevD.52.679].

51. J. Borrill, T.K., T. Vachaspati and A. Vilenkin, Defect production in slow first-order phase transitions, *Phys. Rev. D* **52**, 1934–1943 (1995), Imperial/TP/94-95/18, hep-ph/9503223 [doi:10.1103/PhysRevD.52.1934].

52. A. Yates and T.K., An extension to models for cosmic string formation, *Physics Letters* **B364**, 149–156 (1995), Imperial/TP/94-95/14, hep-ph/9508383 [doi:10.1016/0370-2693(95)01227-3].

53. V.M.H. Ruutu, V.B. Eltsov, A.J. Gill, T.K., M. Krusius, Yu.G. Makhlin, B. Plaçais, G.E. Volovik and Wen Xu, Vortex formation in neutron-irradiated superfluid ^{3}He-B as an analogue of cosmological

defect formation, *Nature* **382**, 334–336 (1996), Imperial/TP/95-96/17, cond-mat/9512117 [doi:10.1038/382334a0].

54. A.J. Gill and T.K., Quench induced vortices in the symmetry broken phase of liquid ^{4}He, *J. Phys. A: Math. & Gen.* **29**, 4289–4305 (1996).

55. T.K. and G.E. Volovik, On phase ordering behind the propagating front of a second-order transition, *Pis'ma Zh. Eksp. Teor. Fiz.* **65**, 96–101 (1996) [*JETP Letters* **65**, 102–107 (1997)] [doi:10.1134/1.567332].

56. T.K., G. Lozano and A. Yates, Non-Abelian string conductivity, *Phys. Rev. D* **56**, 1204–1214 (1997), Imperial/TP/96-97/14, hep-ph/9701240 [doi:10.1103/PhysRevD.56.120].

57. T.S. Evans, T.K. and D.A. Steer, Wick's theorem for non-symmetric normal ordered products and contractions, *J. Math. Phys.* **39**, 5726–5738 (1998), Imperial/TP/97-98/16, hep-ph/9801404 [doi:10.1063/1.532589].

—. R. Lieu, Y. Takahashi and T.K., Gamma-ray burst as vacuum discharge of super-Schwinger electric fields, unpublished (1998), astro-ph/9803072.

58. E.J. Copeland, T.K. and D.A. Steer, The evolution of a network of cosmic string loops, *Phys. Rev. D* **58**, 043508 (1998) (14 pp), Imperial/TP/97-98/31, hep-ph/9803414 [doi:10.1103/PhysRevD.58.043508].

—. R. Lieu, Y. Takahashi, T.K., J. van Paradijs and A.G. Emslie, Surface quantum effects in a fireball model of gamma ray bursts, astro-ph/9902088 (revised version of astro-ph/9803072), unpublished.

59. T.K. and N.F. Lepora, Classifying vortex solutions to gauge theories, *Phys. Rev. D* **59**, 125019 (1999) (10 pp), Imperial/TP/97-98/48, hep-th/9904177.

60. N.F. Lepora and T.K., Electroweak vacuum geometry, *J.H.E.P.* **9904**, 027 (1999) (19 pp), Imperial/TP/98-99/26, hep-th/9904178.

61. A.C. Davis, T.K., M. Pickles and D.A. Steer, Dynamics and properties of chiral cosmic strings in Minkowski space, *Phys. Rev. D* **62**, 083516 (2000) (8 pp), Imperial/TP/99-0/27, astro-ph/0005514 [doi:10.1103/PhysRevD.62.083516].

62. V.B. Eltsov, T.K., M. Krusius, V.M.H. Ruutu and G.E. Volovik, Composite defect extends analogy between cosmology and ^{3}He, *Phys. Rev. Letters* **85**, 4739 (2000), Imperial/TP/99-0/42, cond-mat/0007369 [doi:10.1103/PhysRevLett.85.4739].

63. A.C. Davis, T.K., A. Rajantie and H. Shanahan, Topological defects in lattice gauge theories, *J.H.E.P.* **11**, 010 (2000) (23 pp), Imperial/TP/99-0/43, hep-lat/0009037.

64. J. Magueijo, H. Sandvik and T.K., Nielsen–Olesen vortex in varying-alpha theories, *Phys. Rev. D* **64**, 023521 (2001) (7 pp), Imperial/TP/ 00-01/07, hep-ph/0101155 [doi:10.1103/PhysRevD.64.023521].

65. A.C. Davis, A. Hart, TK and A. Rajantie, Monopole mass in the three-dimensional Georgi–Glashow model, *Phys. Rev. D* **65**, 125008 (2002) (12 pp), Imperial/TP/01-02/001, hep-lat/0110154 [doi:10.1103/PhysRevD.65.125008].

66. T.K. and A. Rajantie, Estimation of vortex density after superconducting film quench, *Phys. Rev. B* **68**, 174512 (2003) (6 pp), Imperial/TP/02-03/026, cond-mat/0306633 [doi:10.1103/PhysRevB. 68.174512].

67. M.A. Donaire, T.K. and A. Rajantie, Spontaneous vortex formation on a superconductor film, *New J. Phys.* **9**, 148 (2007) (9 pp) (also in *IoP Select*), Imperial/TP/040905, cond-mat/0409172 [doi:10.1088/ 1367-2630/9/5/148].

68. T.K. and R. Lieu, Average magnification effect of clumping of matter, *Astrophys. J.* **632**, 718–726 (2005), Imperial/TP/041203, astro-ph/0412275 [doi:10.1086/444343].

69. E.J. Copeland, T.K. and D.A. Steer, Collisions of strings with Y junctions, *Phys. Rev. Letters* **97**, 021602 (2006) (4 pp), Imperial/TP/ 06/TK/01, hep-th/0601153 [doi:10.1103/PhysRevLett.97.021602].

70. E.J. Copeland, T.K. and D.A. Steer, Constraints on string networks with junctions, *Phys. Rev. D* **75**, 065024 (2007), Imperial/TP/06/ TK/02, hep-th/0611243 [doi:10.1103/PhysRevD.75.065024].

71. P. Salmi, A. Achúcarro, E.J. Copeland, T.K., R. de Putter and D.A. Steer, Kinematic constraints on the formation of bound states of cosmic strings — field-theoretical approach, *Phys. Rev. D* **77**, 041701 (2008), Imperial/TP/07/TK/01, arXiv:0712.1204 [hep-th][doi:10.1103/PhysRevD.77.041701].

72. E.J. Copeland, H. Firouzjahi, T.K. and D.A. Steer, Collision of cosmic superstrings, *Phys. Rev. D* **77**, 063521 (2008), Imperial/TP/07/ TK/02, arXiv:0712.0808 [hep-th] [doi:10.1103/PhysRevD.77.063521].

73. Yifung Ng, T.K. and T. Vachaspati, Formation of non-abelian monopoles connected by strings, *Phys. Rev. D* **78**, 046001 (2008), Imperial/TP/08/TK/02, arXiv:0806.0155 [hep-th] [doi:10.1103/Phys-RevD.78.046001].

—. R. Lieu and T.K., A natural origin of primordial density perturbations, Imperial/TP/09/TK/01, arXiv:0904.4840 [astro-ph].

74. E.J. Copeland and T.K., Kinks and small-scale structure on cosmic strings, *Phys. Rev.* D **80**, 123523 (2009), Imperial/TP/09/TK/02, arXiv:0909.1960 [astro-ph] [doi: 10.1103/PhysRevD.80.123523].

75. J. Magueijo, T. Zlosnik and T.K., Cosmology with a spin, *Phys. Rev.* D **87**, 063504 (2013), arXiv:1212.0585 [astro-ph.CO] [doi: 10.1103/PhysRevD.87.063504].

76. R. Lieu and T.K., Large pre-inflationary thermal density perturbations, *Mon. Not. Roy. Astr. Soc.: Letters*, 2013, arXiv:1110.1172 [astro-ph:CO] [doi: 10.1093/mnrasl/slt097].

77. A. del Campo, T.K. and W. Zurek, Causality and non-equilibrium second-order phase transitions in inhomogeneous systems, *J. Phys.: Cond. Matter* **25** (2013) 404210, arXiv:1302.3648 [cond-mat.stat-mech].

78. T.K. and A. Srivastava, Condensed matter analogues of cosmology, *J. Phys.: Condens. Matter* **25** (2013) 400301 [Introduction to special issue of JPCM].

Conference reports

C1. The quantum theory of gravitation, *High energy physics and elementary particles* (Vienna: IAEA,1965), pp. 885–910. [Lectures at the International Centre for Theoretical Physics, Trieste, May–June 1965].

C2. The Goldstone theorem, *Proceedings of the 1967 International Conference on Particles and Fields*, Rochester, New York, Aug. 1967, ed. C.R. Hagen, G.S. Guralnik and V.S. Mathur (New York: Interscience, 1967), pp. 277–304.

C3. Quantum electrodynamics and other fields, *Inaugural lectures* (London: Imperial College, 1971).

C4. Relativistic non-linear quantum mechanics, *Proceedings of the 19th International Conference on High-Energy Physics*, Tokyo, Aug. 1978, ed. S. Homma, M. Kawaguchi and H. Miyazawa (Tokyo: Physical Society of Japan, 1979), pp. 527–529.

C5. Some implications of a cosmological phase transition, *Physics Reports* **67C**, 183–99 (1980) [*Common Trends in Particle Physics and Condensed Matter Physics*, Proceedings of the *Les Houches Winter Advanced Study Institute*, Feb. 1980], Imperial/TP/79-80/23 [doi:10.1016/0370-1573(80)90091-5].

C5. Is a semi-classical theory of gravity viable?, *Quantum Gravity 2: A Second Oxford Symposium*, ed. C.J. Isham, R. Penrose and D.W.

Sciama (Oxford: Clarendon Press, 1981), pp. 63–80.

C6. Phase transitions in the early universe, *Quantum structure of space and time*, ed. M.J. Duff and C.J. Isham (Cambridge: Cambridge University Press, 1982), pp. 391–408. [Proceedings of the *Nuffield Workshop*, Imperial College, Aug. 1981].

C7. Monopoles in the present and early universe, *Monopoles in quantum field theory*, ed. N.S Craigie, P. Goddard and W. Nahm (Singapore: World Scientific, 1982), pp. 341–76 [Proceedings of the *Monopole Meeting*, Trieste, Dec. 1981], Imperial/TP/81-82/14.

C8. Phase transitions in the early universe and their consequences, *Phil. Trans. Roy. Soc.* **A310**, 293–302 (1983) [Lecture at Royal Society Discussion Meeting on *The Constants of Physics*, May 1983].

C9. Evolution of a system of cosmic strings, *Nucl. Phys.* **B252**, 227–44 (1985) [Phase transitions in the very early universe, Proceedings of the International Workshop, Bielefeld, June 1984]; *erratum*, *Nucl. Phys.* **B261**, 750 (1985), Imperial/TP/83-84/54 [doi:10.1016/0550-3213(85)90439-0].

C10. Cosmic strings and galaxy formation, *Particles and the universe*, ed. G. Lazarides and Q. Shafi (North-Holland, 1986), pp. 177–88 [Lecture at the *International Symposium on Particles and the Universe*, Aristotle University of Thessaloniki, June 1985], Imperial/TP/84-85/36.

C11. T.K. and N.G. Turok, Cosmic strings and galaxy formation, *Phil. Trans. Roy. Soc.* **A320**, 565–571 (1986) [Lecture at Royal Society Discussion Meeting on the Material Content of the Universe, October1985], Imperial/TP/85-86/06.

C12. Gauge fields, topological defects and cosmology, *Schrödinger: centenary celebration of a polymath*, ed. C.W. Kilmister (Cambridge: Cambridge University Press, 1987), pp. 201–12 [Lecture presented at the Schrödinger centenary meeting, Imperial College, April 1987].

C13. Cosmic strings — an overview, *The formation and evolution of cosmic strings*, ed. G.W. Gibbons, S.W. Hawking and T. Vachaspati (Cambridge: Cambridge University Press, 1990), pp. 3–32 [Lecture presented at the *Cosmic String Workshop*, Cambridge, July 1989].

C14. T.K. and E. Copeland, Evolution of small-scale structure on cosmic strings, *The birth and early evolution of our universe*, ed. J.S. Nilsson, B. Gustafsson and B.-S. Skagerstam (Singapore: World Scientific, 1991), Physica Scripta T36, 153–166 [Proceedings of *Nobel Symposium 79*, held at Gräftävallen, June 1990], Imperial/TP/89-90/27 [doi:10.1088/0031-8949/1991/T36/017].

C15. Genesis of unified gauge theories — personal recollections from Imperial, *Salamfestschrift*, ed. A. Ali, J. Ellis and S. Randjbar-Daemi (Singapore: World Scientific, 1994), pp. 592–603 [Proceedings of *Salamfest*, held in Trieste, March 1993], Imperial/TP/92-93/24.

C16. Phase transitions in the early universe, *Astrophysics and cosmology: the emerging frontier*, ed. B. Sinha and R.K. Moitra (New Delhi: Narosa Publishing House, 1995), pp. 1–19 [Proceedings of *International Conference on Astrophysics & Cosmology (Birth Centenary Celebration of M.N. Saha)*, Calcutta, December 1993].

C17. Evolution of cosmic strings and cosmological implications, *Second Paris Cosmology Colloquium*, ed. N. Sánchez and H. de Vega (Singapore: World Scientific, 1995), pp. 1–25 [Proceedings of *Journée Comologie*, Paris, June 1994].

C18. Phase transitions and topological defects in the early universe, *Austral. J. Phys.* **50**, 697–722 (1997) [Proceedings of 12th Australian Institute of Physics Congress, Hobart, July 1996] [doi:10.1071/P96076].

C19. Testing cosmological defect formation in the laboratory, Proceedings of Rome conference 1997.

C20. Testing cosmological defect formation in the laboratory, Invited lecture at the Second European Conference on Vortex Matter in Superconductors, Crete, 15–25 September 2001, *Physica C* **369**, 87–92 (2002), Imperial/TP/01-02/005, cond-mat/0111082.

C21. Cosmic strings reborn? Invited lecture at COSLAB 2004, Ambleside, 10–17 September 2004, Imperial/TP/041001, astro-ph/0410073.

C22. T.K. and G.R. Pickett, Introduction. Cosmology meets condensed matter, *Phil. Trans. Roy. Soc. A* (2008) [Introduction to proceedings of Royal Society Discussion Meeting *Cosmology meets condensed matter*, Jan. 2008]

Review articles

R1. G.S. Guralnik, C.R. Hagen and T.K., Broken symmetries and the Goldstone theorem, *Advances in Particle Physics 2*, ed. R.H. Cool and R.E. Marshak (New York: Interscience, 1968), pp. 567–708.

R2. T.K. and K.S. Stelle, Gauge theories of gravity and supergravity, *Progress in quantum field theory*, ed. H. Ezawa and S. Kamefuchi (Amsterdam: North-Holland, 1986), pp. 57–81 [*Festschrift* for H. Umezawa].

R3. Cosmic strings, *J.J. Giambiagi Festschrift*, ed. H. Falomir, R.E. Gamboa Saravi, P. Leal Ferreira and F.A. Schaposnik (Singapore: World Scientific, 1990), pp. 241–257.

R4. Phase transitions and defects in the early universe *M.N. Saha, Birth Centenary Commemoration Volume*, ed. S.B. Karmohaptro (Calcutta: Saha Institute,1993), pp. 1–21.

R5. M.B. Hindmarsh and T.K., Cosmic strings, *Reports on Progress in Physics* **58**, 477–562 (1995), Imperial/TP/94-95/05, hep-ph/9411342.

R6. Topological defects and their homotopy classification *Encyclopedia of Mathematical Physics*, ed. J.-P. Françoise, G.L. Naber and Tsou S.T. (Oxford: Elsevier, 2006) (ISBN 978-0-1251-2666-3), vol. 5, p. 257.

R7. Symmetry breaking in field theory *Encyclopedia of Mathematical Physics*, ed. J.-P. Françoise, G.L. Naber and Tsou S.T. (Oxford: Elsevier, 2006) (ISBN 978-0-1251-2666-3), vol. 5, p. 198.

R8. Englert-Brout-Higgs-Guralnik-Hagen-Kibble mechanism, *Scholarpedia* **4**(1):6441 (2009).

R9. Englert-Brout-Higgs-Guralnik-Hagen-Kibble mechanism (history) *Scholarpedia* **4**(1):8741 (2009) [doi:10.1142/S0217751X09048137].

R10. E.J. Copeland and T.K., Cosmic strings and superstrings, *Proc. Roy Soc. A* **466**, 623–657 (2010), Imperial/TP/09/TK/03, arXiv: 0911.1345 [hep-th] [doi:10.1098/rspa.2009.0591].

Textbooks

Classical mechanics
> 1st ed. (London: McGraw-Hill, 1966)
> 2nd ed. (London: McGraw-Hill, 1973)
> 3rd ed. (London: Longmans, 1985)
> 4th ed. [T.K. and F.H. Berkshire] (London: Addison-Wesley–Longman, 1996)
> 5th ed. [T.K. and F.H. Berkshire] (London: Imperial College Press, 2004)

Mecânica clássica
> (Portuguese ed.), tr. A.L. da Rocha Barros and D.M. Redondo (São Paulo: Editôra Polígono, 1970)

Méchanique classique
> (French ed.), tr. M. Le Ray and F. Guérin (Paris: Ediscience, 1972)

Mecânica clásica
> (Spanish ed.), tr. A. Madroñero de la Cal (Bilbão: Ediciones Urmo, 1972)

Κλασικη Μηχανικη
(Greek ed.), tr. Π. Δητσας and Δ. Σαρδελης (Ηρακλειο: Πανεπιστημιαες Εκδοσεις Κρητης, 2012)

Summer school lectures

S1. Some applications of coherent states, *Cargèse lectures in physics*, vol. 2, ed. M. Lévy (New York: Gordon and Breach, 1968), pp. 239–345 [Lectures given at *Cargèse Summer Institute*, 1967].

S2. Coherent states and infrared divergences, *Lectures in theoretical physics*, vol. XI–D, ed. K. Mahanthappa and W.E. Brittin (New York: Gordon and Breach, 1969), pp. 387–477 [Lectures given at the *Summer Institute for Theoretical Physics*, Boulder, Colorado, 1968].

S3. Quantum electrodynamics, *Quantum optics*, ed. S.M. Kay and A. Maitland (London: Academic Press, 1970), pp. 11–52 [Lectures given at the *Scottish Universities Summer School*, 1969].

S4. Restoration of broken symmetries, Lectures given at the *International School of Elementary Particle Physics*, Bako Polje Makaraska, 1975, ed. M. Nikoli.

S5. Phase transitions in the early universe, *Acta Physics Polonica* **B13**, 723–46 (1982) [Lectures given at the *VI Autumn School of Theoretical Physics*, Szczyrk, 1981], Imperial/TP/81-82/19.

S6. Cosmic strings, *Cosmology and Particle Physics*, ed. E. Alvarez, R. Domínguez Tenreiro, J.M. Ibánez Cabanell and M. Quirós (Singapore: World Scientific, 1987), pp. 171–208 [Lectures presented at the *GIFT XVII International Seminar on Theoretical Physics*, Peñiscola, June 1986].

S7. Cosmic strings — current status, *Current Topics in Astrofundamental Physics*, ed. N. Sánchez and A. Zichichi (Singapore: World Scientific, 1992) pp. 68–91 [Lecture given at *First Daniel Challonge International School of Astrophysics*, Erice, September 1991], Imperial/TP/91-92/03.

S8. Phase transitions in the early universe and defect formation, *Formation and interactions of topological defects*, ed. A.C. Davis and R. Brandenburger (New York: Plenum, 1995), pp. 1–26 [Lectures at NATO Advanced Study Institute held in Cambridge, August 1994].

S9. Early days of gauge theories — recollections from Imperial College, Lecture at the 20th Nathiagali Summer College on Physics and Contemporary Needs, Bhurban, Murree Hills, Pakistan, June 1995, ed. M. Riazuddin, K. A. Shoaib *et al.* (New York: Nova Science, 1995).

S10. Cosmic strings, Lecture at the 20th Nathiagali Summer College on Physics and Contemporary Needs, Bhurban, Murree Hills, Pakistan, June 1995, ed. M. Riazuddin, K. A. Shoaib *et al.* (New York: Nova Science, 1995).

S11. Formation of defects in the early universe — and in the laboratory, *Current Topics in Astrofundamental Physics*, 5th course, ed. N. Sánchez and A. Zichichi (Singapore: World Scientific, 1997) pp. 322–42 [Lecture given at *Daniel Challonge International School of Astrophysics*, Erice, September 1996].

S12. Classification of topological defects and their relevance to cosmology and elsewhere, *Topological Defects and the Non-Equilibrium Dynamics of Symmetry Breaking Phase Transitions*, ed. Y.M. Bunkov and H. Godfrin, NATO Science Series C: Mathematical and Physical Sciences, 549, pp. 7–31 (Dordrecht: Kluwer Academic Publishers, 2000) [Proceedings of NATO Advanced Study Institute, Les Houches, February 1999].

S13. Symmetry breaking and defects, *Patterns of Symmetry Breaking*, ed. H. Arodz, J. Dziarmaga and W.H. Zurek, NATO Science II, 127, pp. 3–36 [Proceedings of NATO Advanced Study Institute, Cracow, September 2002], Imperial/TP/02-3/5, cond-mat/0211110.

Popular and non-technical articles

P1. Lasers in fundamental physics, *New Scientist*, no. 404, p. 372, 14 Aug 1964.

P2. J.M. Charap and T.K., Excitement in high energy physics, *The Times Science Review*, Spring 1965.

P3. Putting relativity into particle symmetries *The Times Science Review*, Spring 1965.

P4. Elementary particle symmetries, *Contemporary Physics* **6**, 436–52 (1965) [doi:10.1080/0010751650824-7974].

P5. Taxonomy of elementary particles Lecture at *British Association* meeting, Exeter, 1969.

P6. L.S. Brown, L. Castillejo, H.F. Jones, T.K. and M. Rowan-Robinson, Are there real limits to growth? — A reply to Beckerman *Oxford Economic Papers* **25**, 455–60 (1973).

P7. Domains and strings, *Physics Bulletin*, Aug. 1976, p. 337.

P8. Nobel Prizes 1979: Unification, *Physics Bulletin* **30**, 514–15 (1979).

P9. Science and the arms race, *Fire foreloesniger om freds-og konflikt-*

forskning på Københavns Universitet, ed. O. Nathan (1983), pp. 51–71.

P10. Science and the arms race, *A fegyverkezési verseny és a nukleáris háboru*, ed. M. Neményi (Budapest: Hungarian Academy of Sciences, 1984), pp. 31–41 [Proceedings of a meeting in Budapest, March 1984].

P11. Phase transitions: Cosmology in the laboratory, *Nature (News and Views)* **317**, 472 (1985).

P12. Missile defence and European security, *Ways out of the arms race* [*International Scientists' Peace Congress*, Hamburg, November 1986].

P13. Paul Taunton Matthews 1919–1987, *Biographical Memoirs of Fellows of the Royal Society* **34**, 555–580 (1988).

P14. Opening remarks, *Ways out of the arms race: from the nuclear threat to mutual security*, ed. J. Hassard, T.K. and P. Lewis (Singapore: World Scientific, 1989), pp. 1–2 [Proceedings of *Second International Scientists' Congress*, Imperial College, December 1988].

P15. Of pencils, particles and unification, *Physics World*, September 1993, pp. 27–28 [one of the winning entries in a competition to explain the Higgs boson on one side of A4 set by William Waldegrave].

P16. Obituary of Professor Abdus Salam, *The Independent*, 29 November 1996.

P17. Chris Isham and T.K., Abdus Salam 1926–96, *Physics World* **10**, 54 (January 1997).

P18. Muhammad Abdus Salam 1926–1996, *Biographical Memoirs of Fellows of the Royal Society* **44**, 385–401 (1998).

P19. Recollections of Abdus Salam at Imperial College, *The Abdus Salam Memorial Meeting*, ed. J. Ellis, F. Hussain, T.K., G. Thompson and M. Virasoro (Singapore: World Scientific, 1999), pp. 1–11 [Proceedings of the Abdus Salam Memorial Meeting, Trieste, 19–22 November 1997].

P20. Relatively right, *New Scientist* **158**, issue 2137, p. 66.

P21. A.-C. Davis and T.K., Fundamental cosmic strings, *Contemporary Physics* **46**, 313–322 (2005), Imperial/TP/05/TK/01, hep-th/0505050 [doi:10.1080/00107510500165204].

P22. S. Arnott, T.K. and T. Shallice, Maurice Hugh Frederick Wilkins 1916–2004, *Biographical Memoirs of Fellows of the Royal Society* **52**, 455–478 (2006).

P23. Phase transition dynamics in the lab and the universe, *Physics Today* **60**, no. 9 (September 2007), pp. 47–53. [doi:10.1063/1.2784684] [Japanese translation in Parity 2008/03 pp, 4–11.]

P24. Abdus Salam at Imperial College, Proceedings of *Salam + 50*, Conference, Imperial College, July 2007, pp. 10–20 (Singapore: World Scientific, 2008).

Books edited

Ways out of the arms race: from the nuclear threat to mutual security
ed. J. Hassard, T.K. and P. Lewis (Singapore: World Scientific, 1989), [Proceedings of *Second International Scientists' Congress*, Imperial College, December 1988].
Selected papers of Abdus Salam (with commentary)
ed. A. Ali, C. Isham, T.K. and Riazuddin (Singapore: World Scientific, 1994).
The Abdus Salam Memorial Meeting
ed. J. Ellis, F. Hussain, T.K., G. Thompson and M. Virasoro (Singapore: World Scientific, 1999), [Proceedings of the Abdus Salam Memorial Meeting, Trieste, 19–22 November 1997].
Mathematical Physics 2000
ed. A. Fokas, A. Grigoryan, T.K. and B. Zegarlinski (London: Imperial College Press, 2000).
XIIIth International Congress on Mathematical Physics
ed. A. Fokas, A. Grigoryan, T.K. and B. Zegarlinski (Boston: International Press, 2001), [Proceedings of the Congress held at Imperial College, London, 17–22 July, 2000].
Highlights of Mathematical Physics
ed. A. Fokas, J. Halliwell, T.K. and B. Zegarlinski, (Providence, RI: American Mathematical Society, 2002).

Book reviews (selected)

A rebel from the start: Review of Sage: *a life of J.D. Bernal*, by M. Goldsmith, *ICON*, no. 21, Oct. 1980, pp. 17–18.
Review of Selected papers 1945–50, with commentary, by Chen Ning Yang, *Contemp. Phys.* **25**, 407–8 (1984).
The final theory: Review of Dreams of a Final Theory, by S. Weinberg, *Contemp. Phys.* **34**, 99–100 (1993).
Review of Arms and the Physicist, by H.F. York, *Contemp. Phys.* **36**, 359–60 (1995).
Review of Quantum Theory of Fields, vol. 1, by S. Weinberg, *Contemp. Phys.* **37**, 423–4 (1996).

Quantum ghosts get real: Review of Quantum Theory of Fields, vol. 2, by
  S. Weinberg, *New Scientist* **152**, no. 2054, 44–5 (1996).

Review of The Large, the Small and the Human Mind, by R. Penrose, ed.
  M. Longair, *Contemp. Phys.* **39**, 155–6 (1998).

Review of Paul Dirac: the Man and His Work, ed. P. Goddard, *Eur. J.
  Phys.* **19**, 315 (1998).

Review of Physics from Fisher Information by R. Frieden, *Contemp. Phys.*
  40, 357 (1999).

Review of Quantum Aspects of Life, ed. D. Abbott, P.C.W. Davies and
  A.K. Pati, *Int. J. Quantum Information* **8**, 1427-32 (2010) [doi:
  10.1142/S0219749919996039].

www.ingramcontent.com/pod-product-compliance
Lightning Source LLC
Chambersburg PA
CBHW060312220326
41598CB00027B/4310